GENERAL EQUILIBRIUM ANALYSIS

General Equilibrium Analysis

Existence and Optimality Properties of Equilibria

by

Monique Florenzano
Research Director at CNRS, CERMSEM,
Université de Paris 1, France

KLUWER ACADEMIC PUBLISHERS
BOSTON / DORDRECHT / LONDON

A C.I.P. Catalogue record for this book is available from the Library of Congress.

ISBN 1-4020-7512-X

Published by Kluwer Academic Publishers,
P.O. Box 17, 3300 AA Dordrecht, The Netherlands.

Sold and distributed in North, Central and South America
by Kluwer Academic Publishers,
101 Philip Drive, Norwell, MA 02061, U.S.A.

In all other countries, sold and distributed
by Kluwer Academic Publishers,
P.O. Box 322, 3300 AH Dordrecht, The Netherlands.

Printed on acid-free paper

Printed in the Netherlands.

Contents

Preface

This book is a systematic exposition of the Walrasian model of economic equilibrium with a finite number of agents, as formalized by Arrow, Debreu and McKenzie at the beginning of the fifties and since then extensively used, worked and studied. Existence and optimality of general equilibrium are developed repeatedly under different sets of hypothesis which define some general settings. The objective of this monograph is to give to each problem in each framework the most general solution, at least for the present state of art.

As the french monograph I published in 1981, this book gives a large space to a survey of the fixed point theory for correspondences. The renewals in general equilibrium and fixed point theories were for a while intimately linked. It may be useful to go a little bit further than the simple statement of a fixed point result used as a tool in an equilibrium existence proof and to understand the context in which this result is established. Obviously the content of Chapter 1, devoted to fixed points and maximal elements, does not pretend to any exhaustivity. Its organization corresponds to (but voluntarily exceeds) the needs of the following chapters. In order to allow the reader, who is not used in fixed point theory and may want to skip this chapter, to follow without difficulty the subsequent developments, specific fixed point results or arguments are recalled in each of Chapters 2–5, with sometimes an easier proof.

The standard general equilibrium model is introduced at the beginning of Chapter 2. We then focuses on the most familiar framework, at least for economists, of a production economy where it is assumed that each consumer has a complete preorder of preferences on his consumption set. This assumption defines what is called in the monograph "transitive equilibrium". Existence of transitive quasiequilibrium is proved through an excess (quasi-)demand approach, also familiar for economists, though maybe not at the level of generality maintained in the chapter. The last section is devoted to a thorough study of the conditions under which a quasiequilibrium is actually an equilibrium, sufficiently general to be used in all the chapters of the book.

Replacing complete preference preorders by preference correspondences allows for introducing externalities taking the form of a dependence of preferred sets on the actions of other agents, consumers and producers, and also on relative prices. Existence of what is called in the monograph "nontransitive equilibrium" is established in Chapter 3 through a game theoretic approach of simultaneous optimization by the different agents of the economy, including a fictitious auctioneer who sets the prices. Using a fixed point argument due to Gale and Mas-Colell, the equilibrium existence proof follows steps which parallel those of Chapter 2.

Studying the optimality properties of equilibrium in the model, Chapter 4 actually introduces a third equilibrium existence proof. Existence first of weakly Pareto optimal consumption allocations then of Edgeworth equilibria is proved, before studying the conditions of the decentralization by prices of these allocations, so as to get a quasiequilibrium relative to a price system in the first case, a quasiequilibrium of the economy in the second case. Once again, passing from quasiequilibrium to equilibrium uses the results established at the end of the first chapter.

The final chapter is devoted to the extension of the general equilibrium model to economies defined on an infinite dimensional commodity space. Building upon the analysis of Chapter 4, Chapter 5 bases the quasiequilibrium existence in an infinite dimensional setting on a unique existence result for Edgeworth equilibria, followed by several decentralization results. As in the rest of the monograph, we work in this chapter at the highest possible level of generality in the assumptions. We have not tried to simplify the proofs by adopting some strong assumptions. The analysis of economies with an infinite number of commodities is by no way easy. As it stands, I hope that this last chapter will be useful for researchers.

Nonconvex economies and economies with a measure space of agents are out of the scope of this book. Likewise stay out of the scope of this book a number of models: economies with public goods, with money, with dividends, without satiation, with incomplete markets of assets, overlapping generations of agents, ..., which depart from the standard Walrasian model but are studied using essentially the same tools and with the same objectives of generality.

The intended readers of this monograph are senior undergraduate students, graduate students, specialists and researchers in economics, especially in mathematical economics. The mathematical prerequisites are mild for Chapters 2–4 which deal with the finite dimensional model. Nothing more than the usual background of graduate students in economics: topology of $\mathbb{R}^\ell$ and finite dimensional optimization. Chapter 1 and Chapter 5 require a good training in general topology, function analysis and Riesz spaces, areas that are nowadays a part of the necessary background of specialists in mathematical economics. For these domains, the text recalls when they are used some important results or definitions and refer the reader to reference books in each of these areas. An appendix, written by Pascal Gourdel, summarizes the main continuity properties of correspondences.

I wish to thank Cuong Le Van, Jean-Philippe Médecin and Leila Triki for reading the typeset of the manuscript. Special thanks go to Pascal Gourdel and V. Filipe Martins Da Rócha for reading and commenting upon the manuscript, providing numerous corrections. Some parts of the book are the result of a collaboration constant with Pascal, more recent with Filipe. I am also indebted to all the co-authors of my different papers, students as well as senior researchers, and to the friends and colleagues I discussed with in seminars or conferences, especially the ones who have encouraged me to write and publish this book. Finally, I thank CNRS Editions for the permission of using a substantial part of the material contained in a monograph I published in french, in 1981. Obviously, the present monograph is new and is the result of twenty years of research in an exciting area which has not delivered all its secrets.

Monique Florenzano Paris, April 2003

Chapter 1

FIXED POINTS AND MAXIMAL ELEMENTS

In a finite or in an infinite dimensional setting, every general competitive equilibrium existence proof lies on a fixed point theorem. Our aim in this chapter is to give a systematic account of some topics in fixed point theory that lie on the border line of topology and equilibrium theory.

The fixed point theorems which are at the starting point of this chapter are rather old 1909 for the first statement by Brouwer [47] of his famous theorem in the unit-ball of $\mathbb{R}^3$, 1929 [118], 1930 [162] and 1935 [174] for the proof of Brouwer's theorem for a simplex of $\mathbb{R}^\ell$ and its extensions first to nonempty compact convex subsets of a normed space then to nonempty compact convex subsets of a Hausdorff locally convex topological vector space; 1941, 1950 [37], 1952 [71, 97]) for Kakutani's theorem [113] and its corresponding extensions.

However the renewal in the forties of the general equilibrium analysis together with the beginnings of Game Theory gave a formidable impulse for new results, new interpretations, and new developments. In this process in which Mathematical Economics have become an area of application (among other ones) for the fixed point theory, decisive results have been obtained in the same time in both area of fixed points and of their applications. Mathematician economists played sometimes an essential role, even if nowadays the developments they have initiated go much further than the foreseeable needs of applications to equilibrium existence theorems.

We do not intend in this chapter to give an extensive overview of the large literature in the fixed point theory related to Mathematical Economics. We focus on the part of this theory most closely related with the equilibrium theory. Even with this limitation, the number of the references quoted in the text gives an idea of the width of the domain. Our objective is to help the reader to understand the underlying structure of argumentation in the different equilibrium existence proofs we will present in the following chapters. This has guided the choice and the organization of the topics.

The starting point of the first section, actually of the whole chapter, is the celebrated Knaster–Kuratowski–Mazurkiewicz lemma, a result based on combinatorial technique and Sperner's lemma, that we take here as given. Brouwer's theorem is immediately deduced. Kakutani's theorem is proved as a first and relatively easy application of the concept of continuous partition of unity. Its validity is extended to Hausdorff locally convex topological vector spaces using the Schauder projection of a compact set into a finite dimensional linear subspace.

For going from the Brouwer theorem to the Kakutani–Fan theorem, the second section uses a different way, based on an infinite dimensional extension, due to Ky Fan, of the Knaster–Kuratowski–Mazurkiewicz lemma. This gives as a by-product some results, often used in the general equilibrium theory, on the existence of maximal elements for a (preference) relation.

Based on such a result, and still by associating a continuous partition of unity to an open covering of a paracompact space, the third section proves a variant of the Fan coincidence theorem. This theorem which seriously extends the Kakutani–Fan theorem has several applications. A variant of the well-known (in Economics) Debreu–Gale–Nikaido lemma is one of these applications. Zero and fixed point theorems under boundary conditions are an other application, of interest in some equilibrium existence proofs.

An other useful application, presented in Section 1.4, of Ky Fan's coincidence theorem is a proof of the KKMS lemma used in Game Theory for establishing the nonemptiness of the core of a balanced game without side-payments.

The last section is devoted to Michael's selection theorems and to their application for deducing new theorems from the results of the three first sections. The main motivation for this section is the instrumental role of selections for an equilibrium existence lemma due to Gale and Mas-Colell which will be the fundamental tool for proving in Chapters 3 an 4 the equilibrium existence.

The results of this chapter are stated and established in the most general framework in which they have been obtained. To restrict ourselves to Euclidean spaces could lead the reader to confuse the arguments of the proof with peculiarities of the finite dimensional spaces. For General Topology or Topological Vector Spaces, we refer to french versions of Bourbaki's treatise.[1] However, a reader who is not used with these subjects can read most of this chapter replacing by some Euclidean space the (topological vector) spaces used in the text.

A training in finite dimensional convexity is also necessary. Here, we refer the reader to Rockafellar [157] and to our book with C. Le Van [87]. Two facts, specific of the finite dimension, will be used in this chapter. First, the Euclidean topology is the unique Hausdorff vector space topology on a finite dimensional vector space; thus the topology induced on a finite dimensional convex set by a topological vector space in which it is embedded does not vary with the considered ambient space. Related with this fact, any two compact convex sets with the same finite dimension are homeomorphic. This explains why, depending on the technique of proof, fixed point theorems can be stated and proved alternatively for special prototypes of finite dimensional compact convex sets as unit-balls or simplexes.

1.1 From the Knaster–Kuratowski–Mazurkiewicz lemma to the Kakutani–Fan theorem

1.1.1 Preliminary definitions

Let X and Y be sets and 2^Y denote the set of all subset of Y, i.e. the power set of Y. In the following, we will call *correspondence* and we will note $\varphi\colon X \to Y$ a function from X to 2^Y that assigns to each point x of X a subset $\varphi(x)$ called *image* of x, *upper section* of x, or *value* at x of φ. Note that we do not require the image $\varphi(x)$ to be nonempty. A correspondence $\varphi\colon X \to Y$ can be identified with its *graph* $\mathcal{G}_\varphi$, the set $\{(x,y) \in X \times Y : y \in \varphi(x)\}$. If $A \subset X$, the set

$$\varphi(A) := \bigcup_{x \in A} \varphi(x)$$

defines the image of A. For $y \in Y$, the (possibly empty) set $\varphi^{-1}(y) := \{x \in X\colon y \in \varphi(x)\}$ is called *preimage*, *lower section* or *fiber* of y.

Correspondences are also called multi-valued mappings (point-to-set functions, multifunctions), in contrast with single-valued mappings that

[1] English translations are published at Reading, Massachusets: Addison–Wesley for *Topologie générale* and at Berlin: Springer–Verlag for *Espaces vectoriels topologiques*.

are functions from X to Y. In the following, correspondences will be denoted either by capital roman letters or by greek letters.

Assume that $X \subset Y$. If $f: X \to Y$ is a function, a point $x \in X$ such that $f(x) = x$ is called *fixed point of* f. If $X \subset Y$ and $\varphi: X \to Y$ is a correspondence, a point $x \in X$ such that $x \in \varphi(x)$ is said to be a **fixed point of** φ.

Continuity of correspondences. When X and Y are topological spaces, one defines the following continuity notions for a correspondence $\varphi: X \to Y$.

DEFINITION 1.1.1 *A correspondence $\varphi: X \to Y$ is called **upper semicontinuous at** $x^0 \in X$ if for every open set V of Y for which $\varphi(x^0) \subset V$, there exists a neighborhood U of x^0 ($U \in \mathcal{V}(x^0)$) such that*

$$x \in U \Longrightarrow \varphi(x) \subset V.$$

*The correspondence φ is called **upper semicontinuous on** X if φ is upper semicontinuous at every point of X.*

DEFINITION 1.1.2 *A correspondence $\varphi: X \to Y$ is called **lower semicontinuous at** $x^0 \in X$ if for every open set V of Y for which $\varphi(x^0) \cap V \neq \emptyset$, there exists a neighborhood U of x^0 ($U \in \mathcal{V}(x^0)$) such that*

$$x \in U \Longrightarrow \varphi(x) \cap V \neq \emptyset.$$

*The correspondence φ is called **lower semicontinuous on** X if φ is lower semicontinuous at every point of X.*

One readily verifies that correspondences having open preimages (i.e. open fibers) – and more generally correspondences with an open graph in the product space $X \times Y$– are lower semicontinuous.

DEFINITION 1.1.3 *The correspondence φ is called **continuous at** $x^0 \in X$ (resp. **continuous on** X) if it is upper semicontinuous and lower semicontinuous at $x^0 \in X$ (resp. on X).*

DEFINITION 1.1.4 *A correspondence $\varphi: X \to Y$ is called **closed** if its graph $\mathcal{G}_\varphi$ is closed in the product space $X \times Y$.*

The properties of these different continuity notions of a correspondence are classical and established in Berge [27]. They will be studied in detail and proved in the appendix. Mainly if Y is compact, it is equivalent for a correspondence $\varphi: X \to Y$ to be closed or to be upper semicontinuous on X with closed values in Y. Also, if φ is upper semicontinuous with compact values and if $A \subset X$ is compact, then $\varphi(A)$ is a compact subset of Y. A number of other results of least importance will be used in this chapter.

When Y is a Hausdorff topological vector space, other continuity notions will be defined in section 1.3.

Continuous partition of unity.

DEFINITION 1.1.5 *Let X be a topological space that we shall assume to be Hausdorff and $(A_i)_{i \in I}$ a finite family of subsets of X. The family $(f_i)_{i \in I}$ of continuous functions $f_i: X \to \mathbb{R}_+$ is called* ***continuous partition of unity weakly subordinate*** *(resp.* ***subordinate****)* ***to the family*** *$(A_i)_{i \in I}$ provided that the two following conditions hold:*

1 For every $x \in X$, $\sum_{i \in I} f_i(x) = 1$,

2 For each $i \in I$, $f_i(x) = 0$ if $x \notin A_i$ (resp. $\mathrm{supp}(f_i) \subset A_i$)

where $\mathrm{supp}(f_i)$ denotes the closure (in X) of the set of all points $x \in X$ such that $f_i(x) \neq 0$.

If X is compact and if $(U_i)_{i \in J}$ is a finite open covering of X, the existence of a continuous partition of unity subordinate to $(U_i)_{i \in J}$ is a classical result of General Topology. We will use below the existence of a continuous partition of unity weakly subordinate to $(U_i)_{i \in J}$, a weaker result of immediate proof when X is a metric space[2] (e.g. a compact subset of $\mathbb{R}^\ell$).

1.1.2 Fixed points in $\mathbb{R}^\ell$

Let us first recall the statement of the Knaster–Kuratowski–Mazurkiewicz [118] lemma.

LEMMA 1.1.1 *Let $\{a^0, a^1, \ldots, a^p\}$ be $(p+1)$ affinely independent points of $\mathbb{R}^\ell$ and let $\{F^0, F^1, \ldots, F^p\}$ be a collection of closed subsets of $\mathbb{R}^\ell$. Assume that for every subset $\{i_0, i_1, \ldots, i_r\}$ of the set of indices $\{0, 1, \ldots, p\}$,*

[2] If $(a,b) \mapsto \mathrm{dist}(a,b)$ is a distance on X, let for each $i \in J$, $g_i(x) = \mathrm{dist}(x, X \setminus U_i)$ and $f_i(x) = \frac{g_i(x)}{\sum_{j \in J} g_j(x)}$. The family $(f_i)_{i \in J}$ is a continuous partition of unity weakly subordinate to $(U_i)_{i \in J}$.

one has:

$$\mathrm{co}\{a^{i_0}, a^{i_1}, \ldots, a^{i_r}\} \subset \bigcup_{k=0}^{r} F^{i_k}$$

where $\mathrm{co}\{a^{i_0}, a^{i_1}, \ldots, a^{i_r}\}$ *denotes the convex hull of the set* $\{a^{i_0}, a^{i_1}, \ldots, a^{i_r}\}$*, that is the polytope*[3] *with vertices* $a^{i_0}, a^{i_1}, \ldots, a^{i_r}$*. Then*

$$\bigcap_{i=0}^{p} F^{i} \neq \emptyset .$$

Brouwer's fixed point theorem for a continuous single-valued mapping of a p-simplex of $\mathbb{R}^{\ell}$ into itself follows easily.

THEOREM 1.1.1 *Let* S_p *be a p-simplex of* $\mathbb{R}^{\ell}$ *and* $f\colon S_p \to S_p$ *a continuous (single-valued) mapping. Then* f *has a fixed point, i.e. there exists* $\overline{x} \in S_p$ *such that* $f(\overline{x}) = \overline{x}$*.*

Proof. For each $i = 0, 1, \ldots, p$, define

$$F^{i} = \{x \in S_p : \, f_i(x) \leq x_i\}$$

where x_i and $f_i(x)$ denote respectively the ith barycentric coordinates of x and $f(x)$ with respect to the vertices $a^0, a^1, \ldots, a^p$ of S_p. It follows from the continuity of f, together with the continuity of the mapping which gives for any point of the affine subset generated by S_p its i-barycentric coordinate with respect to the points $a^0, a^1, \ldots, a^p$, that each F^i is a closed subset of $\mathbb{R}^{\ell}$. If moreover $\{i_0, i_1, \ldots, i_r\}$ is a subset of the set of indices $\{0, 1, \ldots, p\}$ and if $x \in \mathrm{co}\{a^{i_0}, a^{i_1}, \ldots, a^{i_r}\}$, we claim that $x \in \cap_{k=0}^{r} F^{i_k}$. To see this, assume by contraposition that for every k, $x \notin F^{i_k}$. Thus for every k, $f_{i_k}(x) > x_{i_k}$ and

$$\sum_{k=1}^{r} f_{i_k}(x) > \sum_{k=1}^{r} x_{i_k} = 1.$$

This is a contradiction with $f(x) \in S_p$.

We can now apply the Knaster–Kuratowski–Mazurkiewicz lemma and get that there exists $\overline{x}$ in $\cap_{i=0}^{p} F^{i}$. From $f_i(\overline{x}) \leq \overline{x}_i \; \forall i = 0, 1, \ldots, p$ and from $\sum_{i=0}^{p} f_i(\overline{x}) = \sum_{i=0}^{p} \overline{x}_i = 1$, we deduce that $f(\overline{x}) = \overline{x}$. □

[3]Such a polytope is called *p-simplex* of $\mathbb{R}^{\ell}$.

COROLLARY 1.1.1 (BROUWER) *Let X be a nonempty compact convex subset of $\mathbb{R}^\ell$ and $f: X \to X$ a continuous (single-valued) mapping. Then there exists an $\overline{x} \in X$ such that $f(\overline{x}) = \overline{x}$.*

Proof. It is a known result of finite dimensional convexity that two nonempty compact convex sets of $\mathbb{R}^\ell$ with the same dimension are homeomorphic[4]. Let p be the dimension of X, and $h: S_p \to X$ be an homeomorphism from a p-simplex S_p onto X. Since f is continuous, the mapping $(h^{-1} \circ f \circ h): S_p \to S_p$ is also continuous and has a fixed point $\overline{z}$. Then $\overline{x} = h(\overline{z})$ is a fixed point of f. □

Kakutani's theorem generalizes the existence of a fixed point to closed correspondences with nonempty convex values from a nonempty compact convex subset of $\mathbb{R}^\ell$ into itself. When the compact convex set X is a p-simplex S_p of $\mathbb{R}^\ell$, this theorem can be proved by approximating the correspondence by continuous functions defined by affine linearity from arbitrary values chosen in the images of the vertices of the successive barycentric simplicial subdivisions of S_p. The fixed point result is then extended by homeomorphism to closed correspondences with nonempty convex values from any nonempty compact convex subset of $\mathbb{R}^\ell$ into itself. This approach was the one used by Kakutani [113]. The proof we present here, borrowed from Terkelsen [170], uses the notion of continuous partition of unity.

THEOREM 1.1.2 (KAKUTANI) *If X is a nonempty compact convex subset of $\mathbb{R}^\ell$, then for every closed correspondence $\varphi: X \to X$ with nonempty convex values (equivalently, for every upper semicontinuous correspondence $\varphi: X \to X$ with nonempty closed convex values) there exists an $\overline{x} \in X$ such that $\overline{x} \in \varphi(\overline{x})$.*

Proof. Given $\varepsilon > 0$, let us consider a covering of X by the family of open balls $(B(x, \varepsilon))_{x \in X}$ with center x and radius ε. Since X is compact, there exist $(x^i_\varepsilon)_{i=1}^{r_\varepsilon}$ such that

$$X \subset \bigcup_{i=1}^{r_\varepsilon} B(x^i_\varepsilon, \varepsilon)$$

and a continuous partition of unity $(\alpha^i_\varepsilon)_{i=1}^{r_\varepsilon}$ weakly subordinate to the finite open covering $(X \cap B(x^i_\varepsilon, \varepsilon))_{i=1}^{r_\varepsilon}$ of X. Taking for each i an arbitrary

[4] To see this, one can show that if $p > 0$, every p-dimensional nonempty compact convex subset of $\mathbb{R}^\ell$ is homeomorphic to the unit-ball of $\mathbb{R}^p$.

point $y_\varepsilon^i \in \varphi(x_\varepsilon^i)$, the formula

$$f_\varepsilon(x) = \sum_{i=1}^{r_\varepsilon} \alpha_\varepsilon^i(x) y_\varepsilon^i$$

defines a continuous function $f_\varepsilon \colon X \to X$. Consequently, by Brouwer's fixed point theorem, there exists an $x^\varepsilon \in X$ such that $x^\varepsilon = \sum_{i=1}^{r_\varepsilon} \alpha_\varepsilon^i(x^\varepsilon) y_\varepsilon^i$ with $\alpha_\varepsilon^i(x^\varepsilon) \neq 0 \Longrightarrow x^\varepsilon \in B(x_\varepsilon^i, \varepsilon) \cap X$.

Let now $(\varepsilon_n) \downarrow 0$ be a decreasing sequence of positive real numbers converging to 0. With each ε_n we associate x^{ε_n}, a fixed point of f_{ε_n}. Since X is compact, we can assume without loss of generality that the sequence x^{ε_n} tends to $\overline{x} \in X$. We now claim that $\overline{x}$ is a fixed point of the correspondence φ.

Assume by contraposition that $\overline{x} \notin \varphi(\overline{x})$. Recalling that φ has closed convex values, it follows from the strict separation theorem that there are a $p \in \mathbb{R}^\ell$ and an $\alpha \in \mathbb{R}$ such that

$$p \cdot \overline{x} < \alpha < p \cdot z \quad \forall z \in \varphi(\overline{x}).$$

Moreover, it follows from the upper semicontinuity at $\overline{x}$ of the correspondence φ that for some $\varepsilon > 0$

$$x \in B(\overline{x}, \varepsilon) \Longrightarrow \alpha < p \cdot z \quad \forall z \in \varphi(x).$$

Let n_0 be such that $n > n_0$ implies $\varepsilon_n < \frac{\varepsilon}{2}$ and $\|x^{\varepsilon_n} - \overline{x}\| < \frac{\varepsilon}{2}$. For every $n > n_0$, one has:

$$x^{\varepsilon_n} = \sum_{i=1}^{r_{\varepsilon_n}} \alpha_{\varepsilon_n}^i(x^{\varepsilon_n}) y_{\varepsilon_n}^i \quad \text{with} \quad y_{\varepsilon_n}^i \in \varphi(x_{\varepsilon_n}^i)$$

and

$$\alpha_{\varepsilon_n}^i(x^{\varepsilon_n}) \neq 0 \Longrightarrow x^{\varepsilon_n} \in B(x_{\varepsilon_n}^i, \varepsilon_n) \cap X$$

which implies

$$\|x_{\varepsilon_n}^i - \overline{x}\| \leq \|x_{\varepsilon_n}^i - x^{\varepsilon_n}\| + \|x^{\varepsilon_n} - \overline{x}\| < \varepsilon_n + \frac{\varepsilon}{2} < \varepsilon.$$

It then follows from this inequality and the choice of ε that for every $n > n_0$, $\alpha_{\varepsilon_n}^i(x^{\varepsilon_n}) \neq 0$ implies $\alpha < p \cdot z \;\; \forall z \in \varphi(x_{\varepsilon_n}^i)$ and, in particular, $\alpha < p \cdot y_{\varepsilon_n}^i$; so that for every $n > n_0$, $\alpha < p \cdot x^{\varepsilon_n}$.

Letting n tend to infinity and passing to limit, one gets $\alpha \leq p \cdot \overline{x}$, in contradiction with $p \cdot \overline{x} < \alpha$. □

1.1.3 Fixed points in locally convex topological vector spaces

The Kakutani–Fan theorem[5] extends Kakutani's theorem to correspondences defined on and with values in a nonempty convex compact subset of a Hausdorff locally convex topological vector space. We will deduce here the Kakutani–Fan theorem from a slightly more general result due to Himmelberg[103]. Let us first introduce the following definition:

DEFINITION 1.1.6 *A correspondence $\varphi\colon X \to Y$ is said to be* ***compact*** *if $\varphi(X)$ is contained in a compact subset of the topological space Y. A function $f\colon X \to Y$ is* ***compact*** *if the correspondence $x \to \{f(x)\}$ is compact.*

PROPOSITION 1.1.1 (HIMMELBERG) *Let E be a Hausdorff locally convex topological vector space and X a nonempty convex subset of E. For every compact, upper semicontinuous correspondence $\varphi\colon X \to X$ with nonempty closed convex values, there exists an $\overline{x} \in X$ such that $\overline{x} \in \varphi(\overline{x})$.*

Proof. Let $\mathcal{B}(0)$ be a local base of closed convex and circled neighborhoods of 0. We first claim that for each $V \in \mathcal{B}(0)$, there is some $x_V \in X$ such that $x_V \in \varphi(x_V) + V$. In the following, $\text{int}(V)$ denotes the interior of V.

To prove the claim, let $V \in \mathcal{B}(0)$ and K compact such that $\varphi(X) \subset K \subset X$. As K is compact, there exists $\{y_V^1, \ldots, y_V^{r_V}\}$, a finite subset of K such that $K \subset \bigcup_{i=1}^{r_V}(\{y_V^i\} + \text{int}(V))$, and a continuous partition of unity $(\alpha_V^i)_{i=1}^{r_V}$ weakly subordinate to the covering $\Big(K \cap (\{y_V^i\} + \text{int}(V))\Big)_{i=1}^{r_V}$. Since X is convex, the polytope K_V generated by $\{y_V^1, y_V^2, \ldots, y_V^{r_V}\}$ is contained in X. We now define the function $p_V\colon K \to K_V$

$$p_V(y) = \sum_{i=1}^{r_V} \alpha_i(y) y_V^i$$

and the correspondences $\varphi_V\colon K_V \to K_V$, $\psi_V\colon K_V \to K_V$ by

$$\varphi_V(x) = p_V \circ \varphi(x) \quad \text{and} \quad \psi_V(x) = \text{co}\, \varphi_V(x).$$

By construction, $\varphi_V(x) \subset \varphi(x) + V$. Indeed, if $z \in \varphi_V(x)$ then for some $y \in \varphi(x)$ one has $z = \sum_{i=1}^{r_V} \alpha_V^i(y) y_V^i$ with

$$\alpha_V^i(y) \neq 0 \Longrightarrow y \in \{y_V^i\} + \text{int}(V) \Longrightarrow y_V^i \in \varphi(x) + V$$

[5] See Ky Fan [71] and also Glicksberg [97].

and $z \in \varphi(x) + V$.

On the other hand, the function p_V is clearly continuous. Moreover, since K is compact, the correspondence φ_V, as the composition product of the continuous function p_V and the closed correspondence φ is closed. Since K_V is compact, the correspondence ψ_V is also closed. Finally, ψ_V is nonempty and, by definition, convex valued. The polytope K_V, being finite dimensional, can be embedded in a finite dimensional Euclidean space which induces on K_V the same topology as E. Applying to ψ_V Kakutani's theorem, there exists $x_V \in \psi_V(x_V)$. It follows from the convexity of the values of φ and the convexity of V that $x_V \in \varphi(x_V)+V$, which proves the claim.

We now prove that there exists $\overline{x}$ in X such that $\overline{x} \in \varphi(\overline{x})$. For each $V \in \mathcal{B}(0)$, choose $x_V \in X \cap (\varphi(x_V) + V)$ so that $(x_V - y_V) \in V$ for some $y_V \in \varphi(x_V)$. The family $\mathcal{B}(0)$ is directed by set inclusion.[6] As K is compact, one can assume that the net $(y_V)_{V \in \mathcal{B}(0)}$ converges to $\overline{x} \in K \subset X$. Writing $x_V - \overline{x} = (x_V - y_V) + (y_V - \overline{x})$, one immediately verifies that the net $(x_V)_{V \in \mathcal{B}(0)}$ also converges to $\overline{x}$. That $\overline{x} \in \varphi(\overline{x})$ follows from the closedness of the correspondence φ. □

The Kakutani–Fan theorem is an immediate corollary of Proposition 1.1.1.

COROLLARY 1.1.2 (KAKUTANI–FAN) *Let E be a Hausdorff locally convex topological vector space and X a nonempty convex compact subset of E. For every closed correspondence $\varphi: X \to X$ with nonempty convex values, there exists an $\overline{x} \in X$ such that $\overline{x} \in \varphi(\overline{x})$.*

The proposition 1.1.1 was proved by Himmelberg as a corollary of the Kakutani–Fan theorem. The direct proof given above has some interesting general features. The function p_V, defined in the first part of the proof, generalizes the *Schauder projection* of a compact set into a finite dimensional linear subspace. The first claim in the proof shows for every $V \in \mathcal{B}(0)$ the existence of a *V-fixed point* of the compact correspondence φ. A compactness argument completes the proof, showing that this is equivalent to the existence of a fixed point of φ.

[6] The intersection of two elements of $\mathcal{B}(0)$ contains an element of $\mathcal{B}(0)$ which follows both in the inclusion ordering.

1.2 From the Ky Fan lemma to the Kakutani–Fan theorem

The Knaster-Kuratowski-Mazurkiewicz lemma (Lemma 1.1.1) admits an infinite dimensional version which is particularly suitable for applications in the fixed point theory. Let us first introduce a definition borrowed from Dugundji–Granas [67].

DEFINITION 1.2.1 *Let E be a vector space and X an arbitrary subset of E. A correspondence $F: X \to E$ is said to be* ***KKM*** *provided that*

$$\mathrm{co}\{x^1, x^2, \ldots, x^p\} \subset \bigcup_{i=1}^{p} F(x^i)$$

for each finite subset $\{x^1, x^2, \ldots, x^p\}$ of X.

LEMMA 1.2.1 (KY FAN) *Let E be a topological vector space, X an arbitrary subset of E, and $F: X \to E$ a KKM-correspondence. If all the sets $F(x)$ are closed and one is compact, then*

$$\bigcap_{x \in X} F(x) \neq \emptyset .$$

Proof. Since one of the sets $F(x)$ is compact, it suffices to prove that every finite intersection $\bigcap_{i=1}^{p} F(x^i)$ is nonempty. Given $\{x^1, \ldots, x^p\}$, consider in $\mathbb{R}^\ell$ $(\ell \geq p)$ the $(p-1)$ -simplex S^{p-1} with vertices the p first vectors of the canonical basis : $e^1 = (1, 0, \ldots, 0)$, $e^2 = (0, 1, 0, \ldots, 0)$,..., $e^p = (0, 0, \ldots, 1, 0, \ldots, 0)$. The mapping $h: S^{p-1} \to E$ defined by

$$h(\sum_{i=1}^{p} \alpha_i e^i) = \sum_{i=1}^{p} \alpha_i x^i$$

for $\alpha_i \geq 0$, $\sum_{i=1}^{p} \alpha_i = 1$, is clearly continuous and the sets $G^i = h^{-1}(F(x^i))$ are closed subsets of $\mathbb{R}^\ell$. By hypothesis, for every subset $\{i_0, i_1, \ldots, i_r\}$ of the set of indices $\{0, 1, \ldots, p\}$, one has

$$\mathrm{co}\{e^{i_0}, e^{i_1}, \ldots, e^{i_r}\} \subset \bigcup_{j=0}^{r} G^{i_j} .$$

Indeed, if $z \in \mathrm{co}\{e^{i_0}, e^{i_1}, \ldots, e^{i_r}\}$ then $h(z) \in \mathrm{co}\{x^{i_0}, x^{i_1}, \ldots, x^{i_r}\} \subset \bigcup_{j=1}^{r} F(x^{i_j})$ and thus $z \in h^{-1}(\bigcup_{j=1}^{r} F(x^{i_j})) = \bigcup_{j=1}^{r} h^{-1}(F(x^{i_j}))$. By the

Knaster–Kuratowski–Mazurkiewicz lemma (Lemma 1.1.1), this implies $\bigcap_{i=1}^{p} G^i \neq \emptyset$, thus

$$\bigcap_{i=1}^{p} F(x^i) \neq \emptyset .$$

□

The proof of Lemma 1.2.1 implicitly contains the following result observed by Sonnenschein [169].

COROLLARY 1.2.1 (SONNENSCHEIN) *Let $\{a^0, a^1, \ldots, a^p\}$ be $(p+1)$ (not necessarily affinely independent) points of $\mathbb{R}^\ell$ and let $\{F^0, F^1, \ldots, F^p\}$ be a collection of closed subsets of $\mathbb{R}^\ell$. Assume that for every subset $\{i_0, i_1, \ldots, i_r\}$ of the set of indices $\{0, 1, \ldots, p\}$, one has:*

$$\mathrm{co}(\{a^{i_0}, a^{i_1}, \ldots, a^{i_r}\}) \subset \bigcup_{k=0}^{r} F^{i_k}.$$

Then

$$\bigcap_{i=0}^{p} F^i \neq \emptyset .$$

A simple consequence of Lemma 1.2.1 is the following important result.[7]

PROPOSITION 1.2.1 (BROWDER–FAN) *let E be a Hausdorff topological vector space and X a nonempty compact convex subset of E. Let $\varphi: X \to X$ be a correspondence with nonempty values. Suppose further that φ has open (in X) fibers, i.e. that*

$$\forall y \in X, \quad \varphi^{-1}(y) \text{ is open in } X.$$

Then there exists $\overline{x}$ in X such that $\overline{x} \in \mathrm{co}\,\varphi(\overline{x})$.

Proof. We argue by contradiction, so assume that for every $x \in X$, $x \notin \mathrm{co}\,\varphi(x)$. For each $y \in X$, define $F(y) = X \setminus \varphi^{-1}(y)$, which is a closed subset of the compact set X. Let $\{y^1, \ldots, y^p\}$ be a finite subset of X and $z \in \mathrm{co}\{y^1, \ldots, y^p\}$. We claim that $z \in \bigcup_{i=1}^{p} F(y^i)$. Otherwise, one

[7]The result appeared for the first time in Ky Fan (1961) [72] with a stronger continuity assumption (the correspondence φ was assumed to have an open graph). The version stated here of Proposition 1.2.1 can be found in Browder (1968) [49] and in Ky Fan (1972) [75]. One should observe that, in this result, the topological vector space is not assumed to be locally convex.

would have for every i, $y^i \in \varphi(z)$, thus $z \in \operatorname{co}\varphi(z)$, a contradiction with our hypothesis. Applying now Lemma 1.2.1, we see that $\bigcap_{y\in X} F(y) \neq \emptyset$. Equivalently, there exists $x^* \in X$ such that $\varphi(x^*) = \emptyset$, a contradiction with the assumption that φ has nonempty values. □

In the economic literature, the correspondence $\varphi\colon X \to X$ is often thought of as a *preference correspondence* on X. In this case, given $x \in X$, $\varphi(x)$ denotes the set of the elements of X which are (strictly) preferred to x, and a reasonable assumption on φ is: $\forall x \in X,\ \ x \notin \varphi(x)$, an assumption reinforced in

$$\forall x \in X, \quad x \notin \operatorname{co}\varphi(x) \tag{1.1}$$

when the preference correspondence φ is not assumed to be convex valued.

If for some x^*, $\varphi(x^*) = \emptyset$, then x^* is said to be a *maximal element* of X for the preference relation on X defined by φ. Proposition 1.2.1 can thus be understood as stating either the existence of a fixed point for a correspondence with nonempty convex values and open fibers, or the existence of a maximal element for a preference correspondence satisfying the same continuity property and the convexity-irreflexivity property stated in (1.1).

This duality of interpretations was first expressed by Bergstrom [29] in the following corollary that, following Browder, we will prove directly from Brouwer's theorem.

COROLLARY 1.2.2 (BERGSTROM) *Let E be a Hausdorff topological vector space and X a nonempty compact convex subset of E. Let $\varphi\colon X \to X$ be a correspondence with open (in X) fibers. Then*

1 *either there exists $\overline{x} \in X$ such that $\overline{x} \in \operatorname{co}\varphi(\overline{x})$,*

2 *or there exists $x^* \in X$ such that $\varphi(x^*) = \emptyset$.*

Proof. Assume that φ is nonempty valued. Define the correspondence $\psi\colon X \to X$ by $\psi(x) = \operatorname{co}\varphi(x)$. The correspondence ψ is clearly nonempty valued. On the other hand, it is readily verified that ψ has open (in X) fibers. Indeed, let $y \in X$ and $x \in \psi^{-1}(y)$. It follows that for some p, $y = \sum_{i=1}^{p} \alpha_i y^i$ with each $y^i \in \varphi(x)$, each $\alpha_i \geq 0$, and $\sum_{i=1}^{p} \alpha_i = 1$. Since each $\varphi^{-1}(y^i)$ is open in X, then $\bigcap_{i=1}^{p} \varphi^{-1}(y^i)$ is open in X, contains x, and satisfies $x' \in \bigcap_{i=1}^{p} \varphi^{-1}(y^i) \Longrightarrow y \in \psi(x')$, i.e., $x' \in \psi^{-1}(y)$.

As φ is nonempty valued, the family $(\psi^{-1}(y))_{y\in X}$ is an open covering of X. Since X is compact, there exist $y^i \in X$, $i = 1, \ldots, r$, such that

$$X = \bigcup_{i=1}^{r} \psi^{-1}(y^i).$$

Let $(\alpha^i)_{i=1}^r$ be a continuous partition of unity weakly subordinate to this finite open covering of X. Let K be the polytope generated by the points $y^1, \ldots, y^r$ and $p: K \to K$ the continuous function defined by

$$p(x) = \sum_{i=1}^{r} \alpha^i(x) y^i.$$

Since the topology induced by the Hausdorff vector space topology of E on the affine subset generated by K coincide with its canonical topology, one can apply to p the Brouwer theorem. There exists $\overline{x} \in K$ such that $\overline{x} = p(\overline{x}) = \sum_{i=1}^{r} \alpha^i(\overline{x}) y^i$. From

$$\alpha^i(\overline{x}) \neq 0 \Longrightarrow \overline{x} \in \psi^{-1}(y^i) \Longrightarrow y^i \in \psi(\overline{x}) = \operatorname{co}\varphi(\overline{x})$$

we get $\overline{x} \in \operatorname{co}\varphi(\overline{x})$, which completes the proof. □

Before stating the next corollary, let us introduce some definitions borrowed from Borglin–Keiding [41].

DEFINITION 1.2.2 *Let E be a Hausdorff topological vector space and X a nonempty convex subset of E. A correspondence $\varphi : X \to X$ is said to be* ***KF***[8] *if φ has open fibers (in X) and if for each $x \in X$, $x \notin \operatorname{co}\varphi(x)$. A KF correspondence $\psi_x : X \to X$ is called a* ***KF-majorant of*** *φ* ***at*** *x if there exists some open neighborhood U_x of x such that for every $z \in U_x$, $\varphi(z) \subset \psi_x(z)$. The correspondence $\varphi : X \to X$ is said to be* ***KF-majorized*** *if for each $x \in X$ such that $\varphi(x) \neq \emptyset$, there is a KF-majorant ψ_x of φ at x.*

If X is convex and compact and if there exists some KF correspondence $\psi : X \to X$ such that for every $x \in X$, $\varphi(x) \subset \psi(x)$, it readily follows from Corollary 1.2.2 that there is some $x^* \in X$ such that $\varphi(x^*) = \emptyset$. The same holds true if φ is KF-majorized.

[8] KF is put for Ky Fan.

COROLLARY 1.2.3 (BORGLIN–KEIDING) *Assume that X is a nonempty convex compact of E and that $\varphi : X \to X$ is KF-majorized. Then there exists $x^* \in X$ such that $\varphi(x^*) = \emptyset$.*

Proof. Assume that for every $x \in X$, $\varphi(x) \neq \emptyset$. In view of the previous definition, let for each $x \in X$ a KF-correspondence ψ_x and an open neighborhood U_x of x such that for every $z \in U_x$, $\varphi(z) \subset \psi_x(z)$. The family $(U_x)_{x \in X}$ is an open covering of X which, by compactness of X, contains a finite subcovering $(U_{x_i})_{i \in J}$, J finite. Let $(\alpha_i)_{i \in J}$ be a continuous partition of unity subordinate to $(U_{x_i})_{i \in J}$. The closed sets $F_i = \operatorname{supp} \alpha_i$ form a closed covering of X (that is $X = \bigcup_{i \in J} F_i$) such that $F_i \subset U_{x_i}$.

Define for $i \in J$ the correspondences $\psi_i : X \to X$ by

$$\psi_i(x) = \begin{cases} \psi_{x_i}(x) & \text{if } x \in F_i \\ X & \text{if } x \notin F_i \end{cases}$$

and the correspondence $\psi : X \to X$ by

$$\psi(x) = \bigcap_{i \in J} \psi_i(x).$$

We first claim that each ψ_i has open fibers in X, consequently that ψ has open fibers in X. Indeed, let $y \in X$ and $x \in \psi_i^{-1}(y)$. Either $x \notin F_i$ and $x' \in X \setminus F_i$ implies $y \in X = \psi_i(x')$, that is $x' \in \psi_i^{-1}(y)$; or $x \in F_i$. In this case, $V_i = U_{x_i} \cap \psi_{x_i}^{-1}(y)$ is an open (in X) neighborhood of x such that for $x' \in V_i$, $\psi_i(x') = \psi_{x_i}(x')$ or X whether x' belongs or not to F_i. In both cases, $y \in \psi_i(x')$, that is $x' \in \psi_i^{-1}(y)$.

We next claim that for every $x \in X$, $x \notin \operatorname{co} \psi(x)$, hence that ψ is KF. Indeed, if for some $x \in X$, $x \in \operatorname{co} \psi(x)$, then $x \in \operatorname{co} \psi_i(x)$ for every $i \in J$. Let $i \in J$ be such $x \in F_i$. One gets $x \in \psi_{x_i}(x)$, which obviously contradicts the assumption that ψ_{x_i} is KF.

Finally for every $x \in X$,

$$\varphi(x) \subset \bigcap_{\{i:\ x \in U_i\}} \psi_{x_i}(x) \subset \bigcap_{\{i:\ x \in F_i\}} \psi_{x_i}(x) = \bigcap_{i \in J} \psi_i(x) = \psi(x).$$

In view of Corollary 1.2.2, let x^* be such that $\psi(x^*) = \emptyset$. It follows that $\varphi(x^*) = \emptyset$, in contradiction with our starting hypothesis. □

As noticed by Borglin–Keiding [41], the Kakutani–Fan theorem can be viewed as a corollary of the previous result.

COROLLARY 1.2.4 (KAKUTANI–FAN) *Assume that E is locally convex and that X is a nonempty convex compact subset of E. Then for every upper semicontinuous correspondence $\varphi : X \to X$ with nonempty closed convex values, there exists $\overline{x} \in X$ such that $\overline{x} \in \varphi(\overline{x})$.*

Proof. Assume that for all $x \in X$, $x \notin \varphi(x)$. We will prove that φ is KF-majorized. Indeed, let $x \in X$. Since E is locally convex and $\varphi(x)$ is closed, there exists V_x, a convex open 0-neighborhood, such that $(\{x\}+V_x)\cap(\varphi(x)+V_x) = \emptyset$. Moreover, since φ is upper semicontinuous at x, there exists another open 0-neighborhood V'_x such that $x' \in (\{x\}+V'_x) \Longrightarrow \varphi(x') \subset (\varphi(x) + V_x)$. The correspondence

$$\psi_x(z) = \begin{cases} \varphi(x) + V_x & \text{if } z \in (V_x \cap V'_x) \\ \emptyset & \text{if } z \notin (V_x \cap V'_x) \end{cases}$$

has convex values with $z \notin \psi_x(z)$, an open graph and majorizes $\varphi(z)$ when $z \in (V_x \cap V'_x)$. It thus follows from Corollary 1.2.3 that $x^* \in \varphi(x^*)$ for some $x^* \in X$. But this contradicts the starting hypothesis, which ends the proof. □

1.3 The coincidence theorem and its consequences

1.3.1 More on continuity of correspondences

If X is a Hausdorff topological space and F a Hausdorff topological vector space, one can introduce the two following continuity notions, the first one originating from Ky Fan [73], the second one from Cornet [54].

DEFINITION 1.3.1 *A correspondence $\varphi: X \to F$ is called* ***upper demicontinuous at*** *$x^0 \in X$ if for every open half-space W of F for which $\varphi(x^0) \subset W$, there exists a neighborhood U of x^0 ($U \in \mathcal{V}(x^0)$) such that*

$$x \in U \Longrightarrow \varphi(x) \subset W.$$

The correspondence φ is called ***upper demicontinuous on*** *X if φ is upper demicontinuous at every point of X.*

In other words, φ is upper demicontinuous at x^0 if for every continuous linear functional p on F and any real number $\alpha \in \mathbb{R}$ such that $p(y) < \alpha$ for every $y \in \varphi(x^0)$, there exists a neighborhood U of x^0 ($U \in \mathcal{V}(x^0)$) with $x \in U \Longrightarrow p(y) < \alpha$ for every $y \in \varphi(x)$. Equivalently, for every continuous linear functional p on F and any real number $\beta \in \mathbb{R}$ such that $p(y) > \beta$ for every $y \in \varphi(x^0)$, there exists a neighborhood U of x^0 with $x \in U \Longrightarrow p(y) > \beta$ for every $y \in \varphi(x)$.

DEFINITION 1.3.2 *A correspondence $\varphi: X \to F$ is called* **upper hemicontinuous at** *$x^0 \in X$ if for every continuous linear functional p on F, the support function $x \mapsto \sup_{y\in\varphi(x)} p(y)$ (equivalently $x \mapsto \inf_{y\in\varphi(x)} p(y)$) is upper semicontinuous (resp. lower semicontinuous) at x^0.*

Equivalently, for every continuous linear functional p on F, and any real number $\alpha \in \mathbb{R}$ such that $\sup_{y\in\varphi(x^0)} p(y) < \alpha$ (resp. any real number $\beta \in \mathbb{R}$ such that $\inf_{y\in\varphi(x^0)} p(y) > \beta$), there exists a neighborhood U of x^0 such that

$$x \in U \Longrightarrow \sup_{y\in\varphi(x)} p(y) < \alpha \text{ (resp. } \inf_{y\in\varphi(x)} p(y) > \beta).$$

The correspondence φ is called **upper hemicontinuous on** *X if φ is upper hemicontinuous at every point of X.*

Note that if two correspondences $\varphi : X \to E$ and $\psi : X \to E$ are upper hemicontinuous at $x^0 \in X$, then the same is true for the correspondence $\lambda\varphi + \mu\psi$ (obviously defined by $(\lambda\varphi+\mu\psi)(x) = \lambda\varphi(x)+\mu\psi(x)$), whatever be λ and μ in $\mathbb{R}$. Indeed,

$$\sup_{y\in(\lambda\varphi+\mu\psi)(x)} p(y) = \sup_{y\in\varphi(x)} \lambda p(y) + \sup_{y\in\psi(x)} \mu p(y)$$

and the upper semicontinuity at x^0 of the functions: $x \to \sup_{y\in\varphi(x)} \lambda p(y)$ and $x \to \sup_{y\in\psi(x)} \mu p(y)$ implies the upper semicontinuity of the function $x \to \sup_{y\in(\lambda\varphi+\mu\psi)(x)} p(y)$.

Obviously, the upper semicontinuity of a correspondence implies its upper demicontinuity, and the upper demicontinuity of a correspondence implies its upper hemicontinuity.[9]

Upper demicontinuity and upper hemicontinuity coincide if φ is compact valued.

Finally, assume that F is locally convex and that φ is upper hemicontinuous on X with closed convex values in some fixed compact subset of F. Then, φ is upper semicontinuous on X. The easy proof of this fact is left for the reader. A similar formulation and related facts can be found in [51, Chap. II, § 4] (see also [2, 16.7]).

[9] One should not confuse the upper hemicontinuity above defined with the notion of hemicontinuity, sometimes used by economists, which actually coincides with the upper semicontinuity as defined here in conformity with the mathematical tradition.

1.3.2 Non-separation and Coincidence theorems

Recall that two nonempty subsets A and B of a topological vector space E are said to be *separated by a closed hyperplane* if there exists a nonzero continuous linear functional p and a real number α such that

$$p(x) \leq \alpha \ \forall x \in A \ \text{ and } \ p(y) \geq \alpha \ \forall y \in B.$$

The two nonempty subsets A and B are said to be *strictly separated by the closed hyperplane* $\{z \in E : p(z) = \alpha\}$ if

$$p(x) < \alpha \ \forall x \in A \ \text{ and } \ p(y) > \alpha \ \forall y \in B.$$

The next result is a variant of a theorem of Ky Fan [75, theorem 3] established for demicontinuous correspondences. We get the same conclusion for hemicontinuous correspondences, strengthening the assumptions on the images of correspondences.

PROPOSITION 1.3.1 (NON-SEPARATION THEOREM) *Let X be a nonempty compact convex subset of a Hausdorff topological vector space E and $\varphi : X \to E$, $\psi : X \to E$ two nonempty valued, upper hemicontinuous correspondences of which one has compact values. Assume that:*

- *for every continuous linear functional p on E and any x of $A(X,p) := \{x \in X : p(x) = \max_{y \in X} p(y)\}$, there exist $u \in \varphi(x)$ and $v \in \psi(x)$ such that $p(u) \leq p(v)$.*

Then there exists $\overline{x} \in X$ such that $\varphi(\overline{x})$ and $\psi(\overline{x})$ cannot be strictly separated by any closed hyperplane.

Proof. Assume the contrary. Since either φ or ψ has compact values, for every $x \in X$ there is some nonzero continuous linear functional p_x such that

$$\sup_{y \in \psi(x)} p_x(y) < \inf_{y \in \varphi(x)} p_x(y).$$

For every nonzero continuous linear functional p, define:

$$V(p) := \{x \in X : \sup_{y \in \psi(x)} p(y) < \inf_{y \in \varphi(x)} p(y)\}.$$

It follows from the upper hemicontinuity of φ and ψ that the $(V(p))$ form an open (in X) covering of X. In view of the compactness of X, let $(V(p_i))_{i=1}^r$ be a finite subcovering of X and $(f_i)_{i=1}^r$ a continuous partition of unity weakly subordinate to this covering. We define a correspondence $\xi : X \to X$ by

$$\xi(x) := \{y \in X : \sum_{i=1}^{r} f_i(x) p_i(y - x) > 0\}.$$

The correspondence ξ is convex valued with $x \notin \xi(x)$; it follows from the continuity of the f_i and p_i that ξ has open (in X) fibers. Thus, by Corollary 1.2.2, for some $x^* \in X$, $\xi(x^*) = \emptyset$, that is

$$\sum_{i=1}^{r} f_i(x^*) p_i(y - x^*) \leq 0 \quad \forall y \in X.$$

Letting $p^* = \sum_{i=1}^{r} f_i(x^*) p_i$, one gets $p^*(y - x^*) \leq 0 \; \forall y \in X$, that is $x^* \in A(X, p^*)$. Applying the condition of Proposition 1.3.1, let $u \in \varphi(x^*)$ and $v \in \psi(x^*)$ be such that $p^*(u) \leq p^*(v)$. Note that

$$f_i(x^*) \neq 0 \Longrightarrow x^* \in V(p_i) \Longrightarrow \sup_{y \in \psi(x^*)} p_i(y) < \inf_{y \in \varphi(x^*)} p_i(y).$$

Using

$$\sup_{y \in \psi(x^*)} p^*(y) \leq \sum_{i=1}^{r} f_i(x^*) \sup_{y \in \psi(x^*)} p_i(y)$$

and

$$\inf_{y \in \varphi(x^*)} p^*(y) \geq \sum_{i=1}^{r} f_i(x^*) \inf_{y \in \varphi(x^*)} p_i(y),$$

one gets

$$p^*(v) \leq \sup_{y \in \psi(x^*)} p^*(y) < \inf_{y \in \varphi(x^*)} p^*(y) \leq p^*(u),$$

which gives a contradiction. □

Since, in a Hausdorff locally convex topological vector space, two closed convex nonempty sets with an empty intersection can be separated by a closed hyperplane, provided one of them is compact, the next proposition is an immediate consequence of the previous one.

PROPOSITION 1.3.2 (COINCIDENCE THEOREM) *Let X be a nonempty compact convex subset of a Hausdorff locally convex topological vector space E and $\varphi : X \to E$, $\psi : X \to E$ two nonempty closed convex valued, upper hemicontinuous correspondences of which one has compact values. Assume that:*

- *for every continuous linear functional p on E and any x of $A(X,p) := \{x \in X : p(x) = \max_{y \in X} p(y)\}$, there exist $u \in \varphi(x)$ and $v \in \psi(x)$ such that $p(u) \leq p(v)$.*

Then there exists $\overline{x} \in X$ such that $\varphi(\overline{x}) \cap \psi(\overline{x}) \neq \emptyset$.

The following corollaries are obtained by specializing one of the correspondences φ and ψ.

Let us first introduce two definitions which rephrase the condition of Proposition 1.3.2 when φ (resp. ψ) equals the identity correspondence $I_X(x) = \{x\}$.

DEFINITION 1.3.3 *If X is a nonempty subset of a Hausdorff topological vector space E, the correspondence $\varphi : X \to E$ is said to be* ***inward*** *if for every continuous linear functional p on E and any x of $A(X,p) := \{x \in X : p(x) = \max_{y \in X} p(y)\}$, there exists $y \in \varphi(x)$ such that $p(y) \leq p(x)$.*

In particular, a correspondence φ such that for each x of X, $\varphi(x) \cap X \neq \emptyset$ is inward.

DEFINITION 1.3.4 *If X is a nonempty subset of a Hausdorff topological vector space E, the correspondence $\varphi : X \to E$ is said to be* ***outward*** *if for every continuous linear functional p on E and any x of $A(X,p) := \{x \in X : p(x) = \max_{y \in X} p(y)\}$, there exists $y \in \varphi(x)$ such that $p(y) \geq p(x)$.*

Definitions 1.3.3 and 1.3.4, due to Ky Fan, slightly weaken inwardness and outwardness conditions given by Browder [49] and Halpern [100]. They yield the following corollary:

COROLLARY 1.3.1 *Let X be a nonempty compact convex subset of a Hausdorff locally convex topological vector space E. Every nonempty closed convex valued, upper hemicontinuous correspondence $\varphi : X \to E$ which is either inward or outward has a fixed point $\overline{x} \in \varphi(\overline{x})$.*

Corollary 1.3.1 obviously contains the Kakutani–Fan theorem. Note that, in the statement of Kakutani–Fan's theorem, it would be no use to replace the upper semicontinuity of the correspondence by its upper hemicontinuity. We already noticed that if X is compact, an upper hemicontinuous correspondence with closed values in X is upper semicontinuous.

Assuming in Proposition 1.3.2 that $\psi(x) = \{0\}$ $\forall x \in X$, we get the following result:

COROLLARY 1.3.2 (CORNET [54]) *Let X be a nonempty compact convex subset of a Hausdorff locally convex topological vector space E and $\varphi : X \to E$ a nonempty closed convex valued, upper hemicontinuous correspondence such that*

- *for every continuous linear functional p on E and any x of $A(X,p) := \{x \in X : p(x) = \max_{y \in X} p(y)\}$, there exists $y \in \varphi(x)$ such that $p(y) \leq 0$.*

Then there exists $\overline{x} \in X$ such that $0 \in \varphi(\overline{x})$.

Moreover, let E' be the topological dual of E, equipped with the weak-star topology $\sigma(E', E)$.[10] Letting $\psi(x) = P^0 \ \forall x \in X$, where P^0 is the polar cone in E of a closed convex cone (with vertex 0) $P \subset E'$ (that is $P^0 = \{z \in E : p(z) \leq 0 \ \forall p \in P\}$), one obtains:

COROLLARY 1.3.3 *Let P be a closed convex cone (with vertex 0) of E'. If $\varphi : X \to E$ is a nonempty compact convex valued, upper hemicontinuous correspondence such that*

- *for every continuous linear functional $p \in P$ and any x of $A(X,p) := \{x \in X : p(x) = \max_{y \in X} p(y)\}$, there exists $u \in \varphi(x)$ such that $p(u) \leq 0$,*

then there exists $\overline{x} \in X$ such that $\varphi(\overline{x}) \cap P^0 \neq \emptyset$.

Proof. Let $\sigma(E, E')$ be the weak topology associated with the duality $\langle E, E' \rangle$. The (convex) polar cone P^0 is $\sigma(E, E')$-closed in E, thus closed in E, and the (constant) correspondence ψ, defined by $\psi(x) = P^0 \ \forall x \in X$, is upper hemicontinuous with nonempty closed convex values.

We now prove that φ and ψ satisfy the condition of Proposition 1.3.2. Let $p \in E'$ and $x \in A(X,p)$. If $p \in P$, according the hypothesis made in the present corollary, there exists $u \in \varphi(x)$ such that $p(u) \leq 0$ with $0 \in \psi(x)$. If $p \notin P$, since $P = P^{00}$, then $p \notin P^{00}$ and $p(v') > 0$ for some $v' \in P^0$. For $x \in A(X,p)$ and $u \in \varphi(x)$, it suffices to let $\lambda = \max\{0, \frac{p(u)}{p(v')}\}$ and $v = \lambda v'$ to get $p(u) \leq p(v)$ with $u \in \varphi(x)$ and $v \in \psi(x) = P^0$.

The desired conclusion follows now from Proposition 1.3.2. □

[10] The weak topology $\sigma(E, E')$ is the weakest locally convex topology on E with which all the continuous linear functional $p \in E'$ remain continuous. The weak-star topology $\sigma(E', E)$ is the weakest locally convex topology on E' with which all the linear functionals $p \mapsto p(x)$, $x \in E$, are continuous.

One should notice that Corollary 1.3.2 strengthens the statement which would result from Corollary 1.3.3 by letting $P = E'$. Corollary 1.3.3 could be proved by applying Corollary 1.3.2 to the correspondence φ' defined by $\varphi'(x) = \varphi(x) - P^0$.

If $E = \mathbb{R}^\ell$ (equipped with the Euclidean norm), E' can be identified with E. Let $\overline{B} := \{x \in \mathbb{R}^\ell : \|x\| \leq 1\}$ be the (closed) unit-ball of $\mathbb{R}^\ell$ and P a closed convex cone (with vertex 0) of $\mathbb{R}^\ell$. If $X = \overline{B} \cap P$ then for $p \in P \setminus \{0\}$, $A(X,p) = \{\frac{p}{\|p\|}\}$. In particular, if $X = \overline{B}$ then for $p \in E' \setminus \{0\}$, $A(X,p) = \{\frac{p}{\|p\|}\}$. Noticing that the condition in Lemmas 1.3.2 and 1.3.3 is automatically satisfied for $p = 0$, this condition in the finite dimensional setting can be thought of as a weakening of the Walras law. This is clear in the two following corollaries, where $S := \{x \in \mathbb{R}^\ell : \|x\| \leq 1\}$ denotes the unit-sphere of $\mathbb{R}^\ell$.

COROLLARY 1.3.4 *Let $\zeta : \overline{B} \to \mathbb{R}^\ell$ be a nonempty closed convex valued, upper hemicontinuous correspondence. Assume that*

$$\forall p \in S,\ \exists z \in \zeta(p),\ p \cdot z \leq 0. \tag{1.2}$$

Then there exists $\overline{p} \in \overline{B}$ such that $0 \in \zeta(\overline{p})$.

COROLLARY 1.3.5 *Let P be a closed convex cone (with vertex 0) of $\mathbb{R}^\ell$ and $\zeta: \overline{B} \cap P \to \mathbb{R}^\ell$ a nonempty compact convex valued, upper hemicontinuous correspondence. Assume that*

$$\forall p \in S \cap P,\ \exists z \in \zeta(p),\ p \cdot z \leq 0. \tag{1.3}$$

Then there exists $\overline{p} \in \overline{B} \cap P$ such that $\zeta(\overline{p}) \cap P^0 \neq \emptyset$.

If P is the positive orthant of $\mathbb{R}^\ell$, the previous corollary is an hemicontinuous variant of the celebrated Debreu–Gale–Nikaido lemma as stated in Debreu [58], Gale [91], and Nikaido [145]. If P is any closed convex cone of $\mathbb{R}^\ell$, the previous corollary is an hemicontinuous variant of a stronger result of Debreu [57]. We will see in the next chapter how Corollaries 1.3.5 and 1.3.4 can be used for proving the existence of a quasiequilibrium, with or without disposal, of an economy.

The last consequence of Proposition 1.3.2 is a surjectivity theorem obtained by letting either $\varphi: X \to E$ or $\psi: X \to E$ to be defined by $\varphi(x) = \{y\}$ (resp. $\psi(x) = \{y\}$) for some $y \in X$.

COROLLARY 1.3.6 *Let X be a nonempty compact convex subset of a Hausdorff locally convex topological vector space E and $\varphi : X \to E$ a nonempty closed convex valued, upper hemicontinuous correspondence. If φ verifies one of the two following properties:*

a φ is outward,

b for every $p \in E'$, for any $x \in A(X,p)$, there exists $u \in \varphi(x)$ such that $p(u) \leq \min_{z \in X} p(z)$ (in this case, φ is said to be ***strongly inward***[11]*),*

then for every $y \in X$ there exists $\overline{x} \in X$ such that $y \in \varphi(\overline{x})$.

COROLLARY 1.3.7 *Let X be a nonempty compact convex subset of a Hausdorff locally convex topological vector space E and $\varphi : X \to X$ a nonempty closed convex valued, upper semicontinuous correspondence. If φ verifies one of the two following properties:*

a for every $p \in E'$, for any $x \in A(X,p)$, $\varphi(x) \cap A(X,p) \neq \emptyset$,

b for every $p \in E'$, for any $x \in A(X,p)$, $\varphi(x) \cap A(X,-p) \neq \emptyset$,

then $X = \varphi(X) = \bigcup_{x \in X} \varphi(x)$ (that is φ is surjective).

The first part of Corollary 1.3.7 generalizes a result of Rogalski [158] to which it is equivalent when E is finite dimensional.

COROLLARY 1.3.8 (ROGALSKI) *Let X be a nonempty compact convex subset of a Hausdorff locally convex topological vector space E and $\varphi : X \to X$ a nonempty closed convex valued, upper semicontinuous correspondence. Under the following condition:*

- *For every extreme subset*[12] *F of X and any $x \in F$, $\varphi(x) \cap F \neq \emptyset$,*

the correspondence φ is surjective.

We will end this section by a remark borrowed from Lasry–Robert [123].

REMARK 1.3.1 Lemma 1.1.1 (the Knaster–Kuratowski–Mazurkiewicz lemma which was the starting point of this chapter) can be seen as a consequence of the previous corollary.

---

[11]This definition is due to Cornet [54].

[12]For a definition and the properties of the extreme subsets (or faces) of a finite dimensional convex set, see Rockafellar [157, Part IV, Section 18]; see also Florenzano–Le Van [87, Chap. 3].

Proof. To see this, let us consider $\{a^0, a^1, \ldots, a^p\}$, $(p+1)$ affinely independent points of $\mathbb{R}^\ell$, and $\{F^0, F^1, \ldots, F^p\}$, $(p+1)$ closed subsets of $\mathbb{R}^\ell$. Assume that for every subset J of the set of indices $\{0, 1, \ldots, p\}$, one has: $\text{co}\{a^i\}_{i\in J} \subset \bigcup_{i\in J} F^i$.

If $\bigcap_{i=0}^{p} F^i = \emptyset$, for each x of the p-simplex $S_p = \text{co}\{a^0, a^1, \ldots, a^p\}$, let $I(x)$ be the nonempty set $I(x) = \{i : x \notin F^i\}$. From $S_p \subset \bigcup_{i\in J} F^i$, we deduce $I(x) \neq \{0, 1, \ldots, p\}$. Define the correspondence $\varphi\colon S_p \to S_p$ by

$$\varphi(x) = \bigcap_{i\in I(x)} \text{co}\{a^j\}_{j\neq i} .$$

It is readily seen that φ has nonempty (if $j \notin I(x)$ then $a^j \in \varphi(x)$), closed, convex values, contained in the relative boundary of S_p. Moreover, if $x' \in S_p \setminus \bigcup_{i\in I(x)} F^i$, then $I(x) \subset I(x')$ and $\varphi(x') \subset \varphi(x)$. As $S_p \setminus \bigcup_{i\in I(x)} F^i$ is an open (in S_p) neighborhood of x, this shows that φ is upper semicontinuous. If finally F is an extreme subset of S_p and if $x \in F$, let F_x be the extreme subset of S_p which contains x in its relative interior. This defines a nonempty $J \neq \{0, 1, \ldots, p\}$ such that $F_x = \text{co}\{a^i\}_{i\in J}$. We then have successively $x \in \bigcup_{i\in J} F^i$, thus for some $i \in J$, $i \notin I(x)$, $a^i \in (\varphi(x) \cap F_x) \subset (\varphi(x) \cap F)$

Applying now Corollary 1.3.8, we get that φ is surjective (i.e. $\varphi(S_p) = S_p$), in contradiction with the fact that $\varphi(S_p)$ is contained in the relative boundary of S_p and that the relative interior of S_p is nonempty. □

1.4 An application of the previous results: the KKMS lemma

For establishing the nonemptiness of the core of a balanced non-side-payment game, game theorists use a generalization of the KKM lemma, due to Shapley [168, Theorem 7.3]. We give here of this result a proof, borrowed from Ichiishi [107] and based on the coincidence theorem (Proposition 1.3.2). Let us first introduce some notation.

- $N := \{1, 2, \ldots, n\}$ is a finite set.

- $\mathcal{N} := 2^N \setminus \{\emptyset\}$ is the family of all nonempty subsets of N.
- For $S \in \mathcal{N}$, let χ_S denote the characteristic function of S:

$$\chi_S(i) = \begin{cases} 1 & \text{if } i \in S, \\ 0 & \text{otherwise.} \end{cases}$$

A subfamily $\mathcal{B}$ of $\mathcal{N}$ is called *balanced* if there exists an indexed set $(\lambda_S)_{S \in \mathcal{B}}$ of nonnegative numbers such that $\sum_{S \in \mathcal{B}} \lambda_S \chi_S = \chi_N$, i.e.

$$\sum_{\{S \in \mathcal{B} \,:\, i \in S\}} \lambda_S = 1, \ \forall i \in N.$$

A geometric interpretation will facilitate the proof of the following result and its comparison with the KKM lemma. Take any set $\{a^0, a^1, \ldots, a^n\}$ of affinely independent points of $\mathbb{R}^\ell$. For each $S \in \mathcal{N}$, define A^S to be the convex hull of the points $\{a^i : i \in S\}$ and let $m_S = \frac{1}{|S|} \sum_{i \in S} a^i$ denote their center of gravity. The reader will verify that $\mathcal{B}$ is balanced if and only if $m_N \in \mathrm{co}\{m_S : S \in \mathcal{B}\}$.

LEMMA 1.4.1 (KKMS) *With the previous notations, let $\{a^0, a^1, \ldots, a^n\}$ be $(n+1)$ affinely independent points of $\mathbb{R}^\ell$ and $(F^S)_{S \in \mathcal{N}}$ be a family of closed subsets of A^N. Assume that for each $T \in \mathcal{N}$,*

$$A^T \subset \bigcup_{S \subset T} F^S. \tag{1.4}$$

Then there exists a balanced subfamily $\mathcal{B}$ of $\mathcal{N}$ for which

$$\bigcap_{S \in \mathcal{B}} F^S \neq \emptyset.$$

Proof. For each $x \in A^N$, let $I(x) = \{S \in \mathcal{N} : x \in F^S\}$. It follows from (1.4) that each $I(x)$ is nonempty. Define a correspondence $\psi : A^N \to A^N$ by $\psi(x) := \mathrm{co}\{m_S : S \in I(x)\}$. It is worth noticing that for each $x \in A^N$, there exists U_x, a neighborhood of x in A^N, such that $x' \in U_x$ implies $I(x') \subset I(x)$, thus $\psi(x') \subset \psi(x)$. It follows that ψ is an upper semicontinuous correspondence with nonempty closed convex values. On the other hand, the constant correspondence $\varphi : A^N \to A^N$ defined by $\varphi(x) := \{m_N\}$ is upper semicontinuous with nonempty compact convex values. In order to check that φ and ψ fulfill the hypothesis of Proposition 1.3.2, it remains to verify the boundary condition. To do this, let p be a linear functional on $\mathbb{R}^\ell$. Either p is constant on A^N and

$A(A^N, p) = A^N$, or for some subset $S \subset N$, $S \neq N$, $A(A^N, p) = A^S$. In either case, for $x \in A(A^N, p)$, it follows from (1.4) that for some $T \subset N$ (resp. $T \subset S$), $p(m_N) \leq p(m_T)$ with $m_T \in \psi(x)$. Applying Proposition 1.3.2, we get the existence of $\overline{x}$ such that $m_N \in \psi(\overline{x})$, i.e. $m_N \in \text{co}\{m_S : S \in I(\overline{x})\}$ and

$$\overline{x} \in \bigcap_{S \in I(\overline{x})} F^S.$$

A glance at the geometric interpretation given above shows that the collection $\mathcal{B} = I(\overline{x})$ is balanced. □

REMARK 1.4.1 If $F^S = \emptyset$ for every S for which $|S| \geq 2$, the statement of Lemma 1.4.1 is reduced to the statement of Lemma 1.1.1, provided we notice that in the statement of Lemma 1.1.1 there is no loss of generality in assuming that each F^k, $k = 0, 1, \ldots, p$, is contained in $\text{co}\{a^0, a^1, \ldots, a^p\}$.[13]

The KKM principle has many other useful applications as intersection theorems, variational inequalities, minimax inequalities, complementarity problems, which are of interest for Mathematical Economics or Game Theory. For these topics, the reader is referred to monographs on fixed point theory (as [67], [40]) and to their references.

1.5 Theorems obtained by selection

Let us begin with a definition.

DEFINITION 1.5.1 *Let $\varphi: X \to Y$ be any correspondence. A function $f: X \to Y$ satisfying $f(x) \in \varphi(x)$ for each $x \in X$ is called a* ***selection function*** *of φ.*

This section is devoted to fixed point, maximal element, zero or surjectivity theorems obtained by applying the results of the previous section not directly to the correspondence φ but to a selection function of φ or to a correspondence defined from a selection function of φ.

In order to precise under which conditions such a construction is possible, we come back to the concept of continuous partition of unity. This concept is the basic tool for proving the selection theorems which will be used in this section.

[13] The reader will also verify that the collection of all singletons in $\{a^0, a^1, \ldots, a^p\}$ is the unique balanced collection of singletons.

1.5.1 Paracompact spaces and continuous partitions of unity

Recall first that a topological space X is called **paracompact** if it is Hausdorff and if for every open covering $(U_i)_{i\in I}$ of X there exists an open covering $(V_j)_{j\in J}$ of X which is

- *finer*: for each $j \in J$ there is some $i \in I$ with $V_j \subset U_i$;
- *locally finite*: each point of X has a neighborhood that meets at most finitely many V_j.

The family $(V_j)_{j\in J}$ is said to be an open **locally finite refinement** of the open covering $(U_i)_{i\in I}$.

From this definition, it obviously follows that a compact topological space is paracompact. Every closed subspace of a paracompact topological space is paracompact. An open subspace of a paracompact topological space is not necessarily paracompact.

We now adapt Definition 1.1.5 in order to deal with continuous partitions of unity weakly subordinate (resp. subordinate) to a non necessarily finite covering of X.

DEFINITION 1.5.2 *Let X be a topological space that we shall assume to be Hausdorff and $(A_i)_{i\in I}$ a family of subsets of X. A family $(f_i)_{i\in I}$ of continuous functions $f_i\colon X \to \mathbb{R}_+$ is called* ***continuous partition of unity weakly subordinate*** *(resp.* ***subordinate****)* ***to the family*** *$(A_i)_{i\in I}$ provided that the two following conditions hold:*

1. *For each $x \in X$, only finitely many functions are nonzero and $\sum_{i\in I} f_i(x) = 1$,*
2. *For each $i \in I$, $f_i(x) = 0$ if $x \notin A_i$ (resp. $\operatorname{supp}(f_i) \subset A_i$)*

where we recall that $\operatorname{supp}(f_i)$ *denotes the closure (in X) of the set of all points $x \in X$ such that $f_i(x) \neq 0$.*

The partition is said to be ***locally finite*** *if the family of the supports of f_i is locally finite, that is, if each point of X has a neighborhood that meets at most finitely many* $\operatorname{supp}(f_i)$.

Obviously, if $(f_i)_{i\in I}$ is subordinate to $(A_i)_{i\in I}$, it is also weakly subordinate; the converse implication does not necessarily hold. However, the useful observation here is that if X is a topological space, $(U_i)_{i\in I}$ an open covering of X, $(f_i)_{i\in I}$ a continuous partition of unity weakly subordinate to $(U_i)_{i\in I}$, then there exists on X a locally finite continuous partition of unity subordinate to $(U_i)_{i\in I}$. This observation (see Bourbaki [43, Topologie générale, Chap. IX, § 4, Proposition 3]) is the main tool for proving that a metrizable space is paracompact.

A paracompact topological space has the following property[14] which extends the property above stated for compact spaces, and completely characterizes paracompactness for Hausdorff topological spaces:

For every open covering $(U_i)_{i\in I}$ of a paracompact topological space X, there exists a locally finite continuous partition of unity subordinate to $(U_i)_{i\in I}$.

This means that there exists a family $(f_i)_{i\in I}$ of continuous functions $f_i: X \to \mathbb{R}_+$ such that

1 For every $x \in X$, there is some neighborhood V_x of x and a finite subset J_x of I such that for each $y \in V_x$, $f_i(y) = 0$ if $i \notin J_x$ and $\sum_{i\in I} f_i(y) = \sum_{i\in J_x} f_i(y) = 1$,

2 For each $i \in I$, $\text{supp}(f_i) \subset U_i$.

As a consequence, every open covering $(U_i)_{i\in I}$ of a paracompact topological space X has a locally finite closed refinement.

1.5.2 Michael's selection theorems

In this paragraph, we state and prove three selection theorems, due to Michael [139]. The first one (Proposition 1.5.1), almost immediate, is quoted in [78], [26], [180]. Its proof proceeds along the same lines as the proof of a lemma of Michael [139, Lemma 4.1] (here after Lemma 1.5.1). The second one, Proposition 1.5.2, is one of the main results in [139] (see the comments after Theorem 3.2″). As the correspondence which has a selection function is required to have closed values, it is of a difficult application in the existence of maximal element for preference correspondences and the equilibrium existence problem. The third one

[14]To verify this property, one first establishes the existence of a continuous partition of unity, $(g_j)_{j\in J}$,subordinate to an open locally finite refinement covering $(V_j)_{j\in J}$ of the open covering $(U_i)_{i\in I}$ of X. Then if $\phi: J \to I$ is a map such that $V_j \subset U_{\phi(j)}$ for each $j \in J$, one defines $(f_i)_{i\in I}$ by $f_i = \sum_{\{j:\phi(j)=i\}} g_j$, with the convention $f_i \equiv 0$ if $i \notin \phi(J)$. It is easily verified that $(f_i)_{i\in I}$ is the required locally finite partition.

(Proposition 1.5.3) does not have this drawback. It is a particular case of Michael [139, Theorem 3.1‴] and has been proved to be instrumental in the finite dimensional general equilibrium theory.

PROPOSITION 1.5.1 *If X is paracompact, Y a topological vector space, $\varphi: X \to Y$ a nonempty convex valued correspondence with open fibers, then there exists a continuous function $f: X \to Y$ such that $f(x) \in \varphi(x)$ for every x in X.*

Proof. For $y \in Y$ let $\varphi^{-1}(y) = \{x \in X : y \in \varphi(x)\}$. The collection $(\varphi^{-1}(y))_{y \in Y}$ is an open covering of X. Since X is paracompact, there exists a locally finite partition of unity $(f_y)_{y \in Y}$ subordinate to $(\varphi^{-1}(y))_{y \in Y}$. It is readily verified that the function $f: X \to Y$ defined by

$$f(x) = \sum_{y \in Y} f_y(x) y$$

is continuous. In view of the convexity of the values of φ, it is a continuous selection function of φ. □

The next lemma states the existence of an approximate selection of φ under a weaker continuity property of φ.

LEMMA 1.5.1 *If X is paracompact, Y a topological vector space, $\varphi: X \to Y$ a nonempty convex valued, lower semicontinuous correspondence, and if V is a convex neighborhood of the origin of Y, then there exists a continuous function $f: X \to Y$ such that $f(x) \in (\varphi(x) + V)$ for every x in X.*

Proof. For each $y \in Y$ define

$$U_y = \{x \in X : y \in (\varphi(x) + V)\} = \{x \in X : \varphi(x) \cap (\{y\} - V) \neq \emptyset\}.$$

Since φ is lower semicontinuous and nonempty valued, the collection $(U_y)_{y \in Y}$ form an open covering of X. In view of the paracompactness of X, let $(f_y)_{y \in Y}$ be a locally finite partition of unity subordinate to $(U_y)_{y \in Y}$. As for each $x \in X$, $\varphi(x) + V$ is a convex subset of Y, the function $f: X \to Y$ defined by

$$f(x) = \sum_{y \in Y} f_y(x) y$$

satisfies all our requirements. This completes the proof of the lemma. □

PROPOSITION 1.5.2 *Assume that X is paracompact, Y a metrizable and complete, locally convex topological vector space, and that $\varphi: X \to Y$ is a lower semicontinuous correspondence with nonempty* **closed** *convex values. Then there exists a continuous function $f: X \to Y$ such that $f(x) \in \varphi(x)$ for every x in X.*

Proof. Since Y is locally convex and metrizable, let $(V_n)_{n=1}^{\infty}$ be a base of 0-neighborhoods in Y consisting of symmetric convex sets, such that $V_{n+1} \subset \frac{1}{2^n} V_n$ for all n. We first construct inductively a sequence $(f_n)_{n=1}^{\infty}$ of continuous functions such that for every $x \in X$,

1 $f_n(x) \in (\varphi(x) + V_n) \quad \forall n \geq 1$

2 $f_n(x) \in (\{f_{n-1}(x)\} + 2V_{n-1}) \quad \forall n \geq 2$.

The existence of f_1 follows from Lemma 1.5.1. Assume that $f_1, \ldots, f_k$ have been chosen to satisfy (1) and (2) for $n = 1, \ldots, k$. Let $\varphi_{k+1}: X \to Y$ be the correspondence defined by

$$\varphi_{k+1}(x) = \varphi(x) \cap (\{f_k(x)\} + V_k).$$

Since f_k is continuous, the correspondence $x \mapsto (\{f_k(x)\} + V_k)$ has an open graph. As φ is lower semicontinuous, it follows that φ_{k+1}, being the intersection of a lower semicontinuous correspondence and of an open graph correspondence, is lower semicontinuous. Applying Lemma 1.5.1, we get the existence of f_{k+1} such that for every $x \in X$, $f_{k+1}(x) \in (\varphi_{k+1}(x) + V_{k+1})$. In view of the choice of the base of 0-neighborhoods in Y and the definition of φ_{k+1}, this implies $f_{k+1}(x) \in (\varphi(x) + V_{k+1})$ and $f_{k+1}(x) \in (\{f_k(x)\} + V_k + V_{k+1}) \subset (\{f_k(x)\} + 2V_k)$, as required for the construction of f_{k+1}.

We next observe that by (2), the sequence of continuous functions, $(f_n)_{n=1}^{\infty}$, is uniformly Cauchy. Since Y is complete, the sequence uniformly converges to a continuous function $f: X \to Y$. The correspondence φ being closed valued in Y and the space Y Hausdorff locally convex, it now follows from (1) that $f(x) \in \varphi(x)$ for every $x \in X$. □

LEMMA 1.5.2 *Assume that X is any subset of $\mathbb{R}^\ell$, and that $\varphi: X \to \mathbb{R}^\ell$ is a lower semicontinuous correspondence with nonempty closed convex values. Then there exists a countable collection F of continuous selections of φ such that for every $x \in X$, the set $\{f(x)\}_{f \in F}$ is dense in $\varphi(x)$.*

Proof. Let us first notice that $\mathbb{R}^\ell$ is locally convex, metrizable and complete, and that X equipped with the metric induced by the Euclidean distance is metrizable, thus paracompact.

As well-known, $\mathbb{R}^\ell$ is separable. So let $\{y_j\}_{j=1}^\infty$ be a countable, dense subset of $\mathbb{R}^\ell$ and $(V_k)_{k=1}^\infty$ the sequence of open balls with center 0 and radius $\frac{1}{2^k}$. For each j and k, the set

$$U_{j,k} = \{x \in X : \varphi(x) \cap (\{y_j\} - V_k) \neq \emptyset\}$$

is open in X by lower semicontinuity of φ. As every open set of the metrizable space X, each $U_{j,k}$ is the union of a countable family of closed (in X) subsets of $\mathbb{R}^\ell$, that is

$$U_{j,k} = \bigcup_{i=1}^{\infty} F_{i,j,k}.$$

Define the correspondences $\varphi_{i,j,k}: X \to \mathbb{R}^\ell$ by

$$\varphi_{i,j,k}(x) = \begin{cases} \varphi(x) & \text{if } x \notin F_{i,j,k} \\ \mathrm{cl}(\varphi(x) \cap (\{y_j\} - V_k)) & \text{if } x \in F_{i,j,k} \end{cases}$$

It is easily checked that each $\varphi_{i,j,k}$ is a lower semicontinuous correspondence with nonempty closed convex values. Applying Proposition 1.5.2, there exists a countable collection of functions $f_{i,j,k}: X \to \mathbb{R}^\ell$ of which each one is a continuous selection function of $\varphi_{i,j,k}$.

On one hand, for every x, $f_{i,j,k}(x) \in \varphi(x)$, thus each $f_{i,j,k}$ is a selection function of φ. On the other hand, let $x \in X$ and $y \in \varphi(x)$. In view of the density of $\{y_j\}_{j=1}^\infty$, take j such that $y_j \in y + V_{k+2} \subset \varphi(x) + V_{k+2}$ and i such that $x \in F_{i,j,k+2}$. Then

$$f_{i,j,k+2}(x) \in \varphi_{i,j,k+2}(x) = \mathrm{cl}(\varphi(x) \cap (\{y_j\} - V_{k+2}) \subset \{y_j\} - \mathrm{cl}(V_{k+2})$$

which implies $f_{i,j,k+2}(x) \in \{y_j\} - V_{k+1}$ and $f_{i,j,k+2}(x) \in \{y\} + V_{k+1} + V_{k+2} \subset \{y\} + V_k$. This proves that for every $x \in X$, the set $\{f_{i,j,k}(x)\}$ is dense in $\varphi(x)$ and the proof is complete. □

PROPOSITION 1.5.3 *Assume that X is any subset of $\mathbb{R}^\ell$. Then every lower semicontinuous correspondence $\varphi: X \to \mathbb{R}^\ell$ with nonempty convex values has a continuous selection function.*

Proof. From φ, let us define the correspondence $\psi: X \to \mathbb{R}^\ell$ by $\psi(x) = \mathrm{cl}(\varphi(x))$. Observe that ψ is lower semicontinuous with nonempty closed convex values. Applying the previous lemma, there exists a collection $\{g_i\}_{i=1}^\infty$ of continuous selection functions of ψ such that for every $x \in X$, the set $\{g_i(x)\}_{i=1}^\infty$ is dense in $\psi(x)$. We define successively:

$$f_i(x) = g_1(x) + \frac{g_i(x) - g_1(x)}{\max\{1, \|g_i(x) - g_1(x)\|\}} \quad \text{for all } i$$

$$f(x) = \sum_{i=1}^{\infty} \frac{1}{2^i} f_i(x).$$

We first observe that each f_i is continuous. Since the series defining f converges uniformly in some neighborhood of every $x \in X$, it follows that the function f is continuous.

We next observe that $f(x) \in \psi(x)$ for every $x \in X$. Indeed, by convexity of $\psi(x)$, each $f_i(x)$ is in $\psi(x)$. On the other hand, the distance $\mathrm{dist}(\sum_{i=1}^{n} \frac{1}{2^i} f_i(x), \psi(x))$ tends to 0 when n tends to ∞. Consequently, since $\psi(x)$ is closed, $f(x) \in \psi(x)$.

To complete the proof, we will show that $f(x)$ is in the relative interior $\mathrm{ri}\,\psi(x)$, thus in $\varphi(x)$. Assume by contraposition that $f(x) \notin \mathrm{ri}\,\psi(x)$. Then there exists a face (or extreme subset)[15] F of $\psi(x)$, distinct of $\psi(x)$, such that $f(x) \in \mathrm{ri}\,F$. As previously, one can prove that for any i,

$$\frac{1}{1-\frac{1}{2^i}} \sum_{j \neq i} \frac{1}{2^j} f_j(x) \in \psi(x)$$

so that the point $f(x) = \frac{1}{2^i} f_i(x) + \sum_{j \neq i} \frac{1}{2^j} f_j(x)$, when it is distinct of $f_i(x)$, is an internal point of some closed segment of $\varphi(x)$ one of whose end points is $f_i(x)$. In either case, one gets that for every i, $f_i(x) \in F$. As, by construction, $f_i(x)$ is either an internal point of the segment whose end points are $g_1(x)$ and $g_i(x)$, or else coincide with $g_i(x)$, one likewise gets that for every i, $g_i(x) \in F$. But the set $\{g_i(x)\}_{i=1}^{\infty}$ is dense in $\psi(x)$, and F is closed. This implies $F = \psi(x)$, contrary to the fact that F is distinct of $\psi(x)$. □

1.5.3 Fixed point, zero and surjectivity theorems

As suggested by Hildenbrand and Kirman [105] in their mathematical appendix, and following an idea which goes back to Cellina [52], using selections it is possible to deduce Kakutani's theorem from Brouwer's theorem. It is that we have implicitly done in the proof of Theorem 1.1.2. Given X, a nonempty compact convex subset of $\mathbb{R}^\ell$, let us associate with the correspondence $\varphi: X \to X$ the correspondences $\psi_\varepsilon: X \to X$ defined

[15] For the definition and properties of the faces of a finite dimensional convex set, see [157, § 18] or [87, Chap. 3].

for every $\varepsilon > 0$ and the open ball $B(x, \varepsilon)$ by

$$\psi_\varepsilon(x) = \mathrm{co}\left(\bigcup_{z \in (B(x,\varepsilon) \cap X)} \varphi(z)\right).$$

In view of the hypothesis on φ and its construction, each ψ_ε has nonempty convex values and open fibers in X. It follows from Proposition 1.5.1 that each ψ_ε has a continuous selection function. The functions f_ε constructed in the proof of Theorem 1.1.2 are such selection functions. The end of the proof was for proving, from the assumptions on φ, that a sequence of fixed points of these selection functions converges to a fixed point of φ.

In what follows, we will use the selection theorems for obtaining new results, starting from the different corollaries of Proposition 1.3.2. As in Section 1.3, if X is convex compact subset of a Hausdorff locally convex topological vector space E, E' will denote the topological dual of E and for each $p \in E'$, we will note

$$A(X, p) := \{x \in X : p(x) = \max_{y \in X} p(y)\}.$$

The first result is a fixed point theorem.

PROPOSITION 1.5.4 *Let X be a nonempty convex compact subset of a Hausdorff locally convex topological vector space E and $\varphi: X \to E$ a nonempty convex valued correspondence. If φ satisfies one of the two following conditions:*

a $\forall p \in E', \ \forall x \in A(X,p), \ \forall y \in \varphi(x), \ p(y) \leq p(x),$

b $\forall p \in E', \ \forall x \in A(X,p), \ \forall y \in \varphi(x), \ p(y) \geq p(x),$

then in each one of the following cases:

1 $\forall y \in E$, *$\varphi^{-1}(y)$ is open in X (i.e. φ has open fibers),*

*2 E is metrizable and complete and φ is lower semicontinuous with **closed** values,*

3 $E = \mathbb{R}^\ell$ *and φ is lower semicontinuous,*

φ has a fixed point, that is there exists $\overline{x} \in X$ such that $\overline{x} \in \varphi(\overline{x})$.

Proof. The proof is immediate. In each one of the three cases 1), 2) or 3), the correspondence φ has a continuous selection function f. Whether

φ satisfies Condition a) or Condition 2, f is inward or outward and the conclusion follows from Corollary 1.3.1. □

In the case 1), for a correspondence $\varphi: X \to X \subset E$ with open fibers, the just obtained result is weaker than the statement of the Browder–Fan theorem (Proposition 1.2.1) which does not assume that E is locally convex. In the case 3), for a lower semicontinuous correspondence $\varphi: X \to X \subset \mathbb{R}^\ell$, the obtained result is due to Bergstrom [29] and precises the previously quoted Bergstrom result (Corollary 1.2.2). As already noticed, it can be used as a maximal element existence theorem for a (strict) preference correspondence as well as a fixed point theorem for a nonempty valued correspondence.

COROLLARY 1.5.1 (BERGSTROM) *Let X be a nonempty compact convex subset of $\mathbb{R}^\ell$ and $\varphi: X \to X$ a lower semicontinuous correspondence. Then*

1 either there exists $\overline{x} \in X$ such that $\overline{x} \in \operatorname{co} \varphi(\overline{x})$,

2 or there exists $x^* \in X$ such that $\varphi(x^*) = \emptyset$.

Proof. Let us assume that φ is nonempty valued. Define $\psi: X \to X$ by $\psi(x) = \operatorname{co} \varphi(x)$. One readily verifies that ψ is lower semicontinuous. It follows from Proposition 1.5.4 that ψ has a fixed point, that is that there exists $\overline{x} \in X$ such that $\overline{x} \in \operatorname{co} \varphi(\overline{x})$. □

Moreover, the previous corollary is contained as a particular case in the following result, obtained following an idea of Gale–Mas-Colell [94] and [96] who state only the part 3) of the result.

PROPOSITION 1.5.5 *Let $(E_i)_{i=1}^m$ be a finite family of Hausdorff locally convex topological vector spaces and, for each i, X_i a nonempty compact convex subset of E_i. Let $X = \prod_{i=1}^m X_i$ be the Cartesian product of the X_i and for each i, $\varphi_i: X \to X_i$ a convex valued correspondence. Then, in each one of the following cases:*

1 $\forall i$, E_i is metrizable and $\forall y_i \in E_i$, $\varphi_i^{-1}(y_i)$ is open in X,

*2 $\forall i$, E_i is metrizable and complete and φ_i is lower semicontinuous with **closed** values,*

3 $\forall i$, $E_i = \mathbb{R}^\ell$ and φ_i is lower semicontinuous,

there exists $\overline{x} \in X$ such that for each $i = 1, \ldots, m$,

$$\text{either } \varphi_i(\overline{x}) = \emptyset \text{ or } \overline{x}_i \in \varphi_i(\overline{x}).$$

Proof. For each i, define $U_i := \{x \in X : \varphi_i(x) \neq \emptyset\}$ and denote by $\varphi_{i|_{U_i}}$ the restriction to U_i of the correspondence φ_i. In each one of the three cases 1), 2), 3), the correspondence $\varphi_{i|_{U_i}}$ has a continuous selection function $f_i: U_i \to X_i$. Next, define the correspondences $\psi_i: X \to X_i$ and $\psi: X \to X$ by:

$$\psi_i(x) = \begin{cases} \{f_i(x)\} & \text{if } x \in U_i \\ X_i & \text{if } x \notin U_i \end{cases}$$

and

$$\psi(x) = \prod_{i=1}^{m} \psi_i(x).$$

Noticing that, due to the lower semicontinuity of φ_i, each U_i is open in X, it is easily verified that each ψ_i is upper semicontinuous with nonempty compact convex values. Being a finite product of upper semicontinuous correspondences with nonempty compact convex values, the correspondence ψ is also upper semicontinuous with nonempty compact convex values. Since X is compact, it follows from the Kakutani–Fan theorem (Corollary 1.1.2) that ψ has a fixed point $\overline{x}$. Thus for each i, either $\overline{x} \in U_i$ and $\overline{x}_i = f_i(\overline{x}) \in \varphi_i(\overline{x})$ or $\overline{x} \notin U_i$ and $\varphi_i(\overline{x}) = \emptyset$. □

COROLLARY 1.5.2 *Let $(E_i)_{i=1}^m$ be a finite family of Hausdorff locally convex topological vector spaces and, for each i, X_i a nonempty compact convex subset of E_i. Let $X = \prod_{i=1}^m X_i$ be the Cartesian product of the X_i and for each i, $\varphi_i: X \to X_i$ a correspondence. Then, in each one of the following cases:*

1 $\forall i$, E_i is metrizable and $\forall y_i \in E_i$, $\varphi_i^{-1}(y_i)$ is open in X,

2 $\forall i$, $E_i = \mathbb{R}^\ell$ and φ_i is lower semicontinuous,

there exists $\overline{x} \in X$ such that for each $i = 1, \ldots, m$,

$$\text{either } \varphi_i(\overline{x}) = \emptyset \text{ or } \overline{x}_i \in \operatorname{co}(\varphi_i(\overline{x})).$$

Proof. It suffices to apply Proposition 1.5.5 to the correspondences ψ_i defined by $\psi_i(x) = \operatorname{co}(\varphi_i(x))$. □

The next corollary, due to Shafer–Sonnenschein [166], uses stronger continuity assumptions on the correspondences φ_i. It is a particular case of a result obtained from Kakutani's theorem without using any selection theorem. It is worth giving its direct proof.

COROLLARY 1.5.3 *Let for each $i = 1, \ldots, m$, X_i be a nonempty compact convex subset of $\mathbb{R}^\ell$, and if $X = \prod_{i=1}^m X_i$, $\varphi_i: X \to X_i$ a correspondence with an open graph in $X \times X_i$. Then there exists $\overline{x} \in X$ such that for each $i = 1, \ldots, m$,*

$$\text{either } \varphi_i(\overline{x}) = \emptyset \text{ or } \overline{x}_i \in \text{co}(\varphi_i(\overline{x})).$$

Proof. Let for each i, $\mathcal{G}_{\varphi_i} = \{(y, x_i) \in X \times X_i : x_i \in \varphi_i(x)\}$ denote the graph of φ_i and $G_i = (X \times X_i) \setminus \mathcal{G}_{\varphi_i}$ its complement in $X \times X_i$. For each i, define the correspondence $\psi_i: X \to X_i$ and $\chi_i: X \to X_i$ by

$$\psi_i(y) = \{x_i \in X_i : \text{dist}((y, x_i), G_i) \geq \text{dist}((y, z_i), G_i), \quad \forall z_i \in X_i\}$$

and

$$\chi_i(y) = \text{co}(\psi_i(y)).$$

The reader will verify that the correspondences χ_i have a closed graph and nonempty convex values. The same holds true for the correspondence $\chi: X \to X$ defined by

$$\chi(x) = \prod_{i=1}^{m} \chi_i(x).$$

Applying Kakutani's theorem (Theorem 1.1.2), there exists $\overline{x} \in X$ such that for each i, $\overline{x}_i \in \chi_i(\overline{x})$.

Noticing that $\text{dist}((y, z_i), G_i) > 0$ if and only if $z_i \in \varphi_i(y)$, we obtain that if $\varphi_i(\overline{x}) \neq \emptyset$ then $\psi_i(\overline{x}) \subset \varphi_i(\overline{x})$, which yields the conclusion of the corollary. □

An infinite dimensional version of the previous corollary can be obtained with the same proof:

COROLLARY 1.5.4 *Let for each $i = 1, \ldots, m$, X_i be a nonempty compact convex subset of a locally convex metrizable topological vector space E_i, and if $X = \prod_{i=1}^m X_i$, $\varphi_i: X \to X_i$ a correspondence with an open graph in $X \times X_i$. Then there exists $\overline{x} \in X$ such that for each $i = 1, \ldots, m$,*

$$\text{either } \varphi_i(\overline{x}) = \emptyset \text{ or } \overline{x}_i \in \text{co}(\varphi_i(\overline{x})).$$

To end with fixed point theorems, it is worth noticing that Kakutani's theorem can be obtained as a consequence of Corollary 1.5.3.

COROLLARY 1.5.5 (KAKUTANI) *If X is a nonempty compact convex subset of $\mathbb{R}^\ell$, then for every closed correspondence $\varphi: X \to X$ with* *nonempty convex values (equivalently, for every upper semicontinuous correspondence $\varphi: X \to X$ with nonempty closed convex values) there exists an $\overline{x} \in X$ such that $\overline{x} \in \varphi(\overline{x})$.*

Proof. Assume on the contrary that φ has no fixed point. It is routine to show that for every $x \in X$ there exist an open neighborhood U_x of x in X and an open convex set V_x such that $U_x \cap V_x = \emptyset$ and $x' \in U_x \Longrightarrow \varphi(x') \subset V_x$. The collection $(U_x)_{x\in X}$ form an open covering of X. Let, since X is compact, $(U_{x_i})_{i=1}^n$ a finite covering of X and $(f_i)_{i=1}^n$ a continuous partition of unity subordinate to this covering. The supports $F_i = \mathrm{supp}(f_i) = \mathrm{cl}(\{x \in X : f_i(x) > 0\})$ of each f_i form a closed covering of X such that $F_i \subset U_{x_i}$ for each $i = 1, \ldots, n$.

Define the correspondence $\psi: X \to X$ by:

$$\psi(x) = \bigcap_{\{i \,:\, x\in F_i\}} V_{x_i}.$$

Note that since for every $x \in X$, $\varphi(x) \subset \psi(x)$, ψ has nonempty and, by construction, convex values.

We first claim that ψ has an open graph. Indeed, if $(x, y) \in \mathcal{G}(\psi)$ let $U = \bigcap_{\{i \,:\, x\notin F_i\}} (X \setminus F_i)$. Then, recalling that for each i, $F_i \subset U_{x_i}$, one easily sees that $x' \in U \Longrightarrow \psi(x) \subset \psi(x')$. Thus $x' \in U$ and $y' \in \psi(x)$ imply $y' \in \psi(x')$. As ψ is by construction open valued, this shows that ψ has an open graph.

The correspondence ψ has obviously no fixed point. Indeed, $x \in F_i$ implies $x \in U_{x_i}$ implies $x \notin V_{x_i}$. This is in contradiction with Corollary 1.5.3. □

The remainder of this section is devoted to variants of Corollaries 1.3.3 and 1.3.7. Their proofs, similar to the previous ones, are left for the reader. Recall that for X compact contained in E and $p \in E'$, $A(X, p) := \{x \in X : p(x) = \max_{y\in X} p(y)\}$.

PROPOSITION 1.5.6 *Let X be a nonempty compact convex of a Hausdorff locally convex topological space, P a $\sigma(E', E)$-closed convex cone with vertex 0, and P^0 the polar cone of P in E, $\varphi: X \to E$ a nonempty convex valued correspondence. Then, under the condition on φ*

$$\forall p \in P,\ \forall x \in A(X, p),\ \forall y \in \varphi(x),\ p(y) \leq 0$$

and in each one of the following cases:

1 $\forall y \in E$, $\varphi^{-1}(y)$ *is open in* X *(i.e.* φ *has open fibers),*

2 E *is metrizable and complete and* φ *is lower semicontinuous with* **closed** *values,*

3 $E = \mathbb{R}^\ell$ *and* φ *is lower semicontinuous,*

there exists $\overline{x} \in X$ *such that* $\varphi(\overline{x}) \cap P^0 \neq \emptyset$.

When X is the closed unit-ball $\overline{B}$ of $\mathbb{R}^\ell$, and if S denotes the unit-sphere, one has in particular a lower semicontinuous variant of the Debreu–Gale–Nikaido lemma.

COROLLARY 1.5.6 *Let* P *be a closed convex cone (with vertex* 0*) of* $\mathbb{R}^\ell$ *and* $\zeta: \overline{B} \cap P \to \mathbb{R}^\ell$ *a nonempty convex valued, lower semicontinuous correspondence. Assume that*

$$\forall p \in S \cap P,\ \forall z \in \zeta(p),\ p \cdot z \leq 0. \tag{1.5}$$

Then there exists $\overline{p} \in \overline{B} \cap P$ *such that* $\zeta(\overline{p}) \cap P^0 \neq \emptyset$.

PROPOSITION 1.5.7 *Let* X *be a nonempty compact convex subset of a Hausdorff locally convex topological vector space* E *and* $\varphi : X \to X$ *a nonempty convex valued correspondence. If* φ *verifies one of the two following properties:*

a for every $p \in E'$, *for any* $x \in A(X,p)$, $\varphi(x) \subset A(X,p)$,

b for every $p \in E'$, *for any* $x \in A(X,p)$, $\varphi(x) \subset A(X,-p)$,

then, in each one of the following cases:

1 $\forall y \in E$, $\varphi^{-1}(y)$ *is open in* X *(i.e.* φ *has open fibers),*

2 E *is metrizable and complete and* φ *is lower semicontinuous with* **closed** *values,*

3 $E = \mathbb{R}^\ell$ *and* φ *is lower semicontinuous,*

$X = \varphi(X) = \bigcup_{x \in X} \varphi(x)$ *(that is* φ *is surjective).*

Chapter 2

TRANSITIVE EQUILIBRIUM OF A FINITE DIMENSIONAL COMPETITIVE ECONOMY

The Walrasian model of general equilibrium of prices, production and exchange is today formulated in the concept of equilibrium of a so-called *private ownership economy.* In the classical model, only a finite number of commodities are exchanged, produced or consumed. Taking prices as given, finitely many consumers (who are endowed with initial holdings of the different commodities and who collect given shares of the profits from production), consume the commodities available on the market by optimizing their preferences among all possible consumption plans which satisfy their resulting budget constraint. Finitely many producers, who also take prices as given, maximize their profit on their individual set of possible production plans. They produce the commodities which satisfy consumers' demand in competition with the initial resources not used by production. Market clearing determines equilibrium prices and the quantities actually consumed and produced at a state of equilibrium.

If one wishes to admit the possibility that, in a state of equilibrium, supply of various commodities exceeds demand, one can add to this definition of the economy the description of possible disposal activities. Thus a classical **private ownership economy** is completely specified by

$$\mathcal{E} = \Big(\langle \mathbb{R}^\ell, \mathbb{R}^\ell \rangle, (X_i, P_i, \omega_i)_{i \in I}, (Y_j)_{j \in J}, (\theta_{i,j})_{\substack{i \in I \\ j \in J}}, Z \Big)$$

with the following interpretation.

- A commodity (physical good or service) is thought of as completely described not only by its characteristics but also by the date, the location and eventually the event on the occurrence of which it is available. The hypothesis of a finite number of commodities, maintained in the current chapter and the two next ones, implies that the economic activity extends over only finitely many possible dates, locations and events. The positive integer ℓ being the number of commodities, the *commodity space* is $\mathbb{R}^\ell$, considered with its linear structure. An element $x \in \mathbb{R}^\ell$ is a *commodity bundle*, that is a vector of zero, positive or negative quantities of each commodity. If $p = (p^h)_{h=1}^{\ell} \in \mathbb{R}^\ell$ is a *price vector* or *price system*, that is a ℓ-list of prices for a unit of each commodity h, denominated in units of account, the *value* of a commodity bundle x relative to the price system p is the inner product $p \cdot x = \sum_{h=1}^{\ell} p^h x^h$. As well-known, each $p \in \mathbb{R}^\ell$ defines a linear functional on $\mathbb{R}^\ell$ under the previous evaluation relation, so that $\langle \mathbb{R}^\ell, \mathbb{R}^\ell \rangle$ is the commodity–price duality of the economy $\mathcal{E}$.

- I is a finite set of consumers. A *consumer* is typically an individual, it may be an household or a larger group with a common purpose. For each $i \in I$,
  - $X_i \subset \mathbb{R}^\ell$ is the *consumption set* of consumer i; this set is interpreted as the set of all commodity bundles with which he can physically survive.[1] A consumption vector $x_i \in X_i$ is a ℓ-list $(x_i^h)_{h=1}^{\ell}$ of the quantities of the various commodities that the ith consumer demands ($x_i^h > 0$) or supplies ($x_i^h < 0$). If $x_i^h = 0$, his net demand (resp. supply) of commodity h is zero.
  - Let $S = \{p \in \mathbb{R}^\ell : \|p\| = 1\}$ be the set of normalized prices. The correspondence $P_i : \prod_{k \in I} X_k \times \prod_{j \in J} Y_j \times S \to X_i$ (*preference correspondence* of i) describes the tastes of consumer i. Under the condition that $x_i \notin P_i(x, y, p)$, $P_i(x, y, p)$ is interpreted as the set of consumption vectors $x'_i \in X_i$ strictly preferred by i to x_i when the consumption vectors of all consumers $k \neq i$ are x_k, the production vectors of all producers j are y_j, the normalized price vector is p. In other words, it is assumed that consumers have preferences over their consumption set with regard to the actions of other agents and to relative prices for all the usual reasons of externalities.
  - $\omega_i \in \mathbb{R}^\ell$ denotes for consumer i an *initial endowment* in each one of the commodities.

[1]Even if the definition of "physical needs", expressed by X_i, is clearly socially determined.

- J is a finite set of producers. The role of a *producer* is to carry out a production plan. For each $j \in J$,

  – $Y_j \subset \mathbb{R}^\ell$ is the (nonempty) *production set* of producer j. It is interpreted as the set of production plans technically possible for the jth producer. A production plan $y_j \in Y_j$ is a ℓ-list $(y_j^h)_{h=1}^\ell$ of the quantities of the various commodities that the jth producer uses ($y_j^h < 0$) or produces ($y_j^h > 0$), that is a specification of all his inputs and all his outputs. If $y_j^h = 0$, his net output (resp. input) of commodity h is zero.

- For all $i \in I$ and $j \in J$, $\theta_{i,j}$ is the contractual claim of the consumer i on the profit of the producer j. The $\theta_{i,j}$ are nonnegative and for every $j \in J$, $\sum_{i \in I} \theta_{i,j} = 1$. Given a price vector p and the production plans $(y_j)_{j \in J}$ of the different producers, the *profit shares* $\theta_{i,j}$ define together with the initial endowments ω_i the individual wealth of each consumer:
$$w_i(p, (y_j)_{j \in J}) = p \cdot \omega_i + \sum_{j \in J} \theta_{i,j} p \cdot y_j.$$
These relations characterize the wealth distribution in a private ownership economy.

- Z (*disposal cone*) is a convex cone with vertex $0 \in Z$, contained in the negative orthant $-\mathbb{R}_+^\ell$ of the Euclidean space $\mathbb{R}^\ell$.

A particular case of private ownership economy is the **exchange economy** where there is no production, specified by:
$$\mathcal{E} = \Big(\langle \mathbb{R}^\ell, \mathbb{R}^\ell \rangle, (X_i, P_i, \omega_i)_{i \in I}, Z\Big).$$

DEFINITION 2.0.3 *An **allocation** $(x, y) \in \prod_{i \in I} X_i \times \prod_{j \in J} Y_j$ is called **feasible** (or **attainable**) if*
$$\sum_{i \in I} x_i - \sum_{j \in J} y_j - \sum_{i \in I} \omega_i \in Z \tag{2.1}$$

DEFINITION 2.0.4 *A t-uple $(\overline{x}, \overline{y}, \overline{p})$ consisting of an attainable allocation $(\overline{x}, \overline{y})$ and a nonzero price vector $\overline{p}$ is said to be an **equilibrium** if*

1 For each $i \in I$, $\overline{p} \cdot \overline{x}_i \leq \overline{p} \cdot \omega_i + \sum_{j \in J} \theta_{i,j} \overline{p} \cdot \overline{y}_j$ and

$$x_i \in P_i(\overline{x}, \overline{y}, \overline{p}) \Longrightarrow \overline{p} \cdot x_i > \overline{p} \cdot \overline{x}_i,$$

2 For each $j \in J$, for all $y_j \in Y_j$, $\overline{p} \cdot y_j \leq \overline{p} \cdot \overline{y}_j$,

*3 $\overline{p} \in Z^0 := \{p \in \mathbb{R}^\ell \,:\, p \cdot z \leq 0, \ \forall z \in Z\}$
where Z^0 is the polar cone of Z, and*

$$\overline{p} \cdot \sum_{i \in I} \overline{x}_i = \overline{p} \cdot \sum_{j \in J} \overline{y}_j + \overline{p} \cdot \sum_{i \in I} \omega_i.$$

Consequently, for every $i \in I$, $\overline{p} \cdot \overline{x}_i = \overline{p} \cdot \omega_i + \sum_{j \in J} \theta_{i,j} \overline{p} \cdot \overline{y}_j$.

Condition (1) states that every consumer has chosen in his consumption set a consumption which best satisfies his preferences under his budget constraint. In view of Condition (3), at equilibrium, each such optimizing consumer necessarily satisfies with equality his budget constraint.
Condition (2) states that every producer has maximized his profit in his production set.
Condition (3) ensures that

$$\overline{z} := \sum_{i \in I} \overline{x}_i - \sum_{j \in J} \overline{y}_j - \sum_{i \in I} \omega_i$$

minimizes on Z the cost of the disposal $(-\overline{z})$ needed for achieving equilibrium and that this cost is equal to zero.

DEFINITION 2.0.5 *A **quasiequilibrium** is a t-uple $(\overline{x}, \overline{y}, \overline{p})$ consisting of an attainable allocation $(\overline{x}, \overline{y})$ and a nonzero price vector $\overline{p}$ which satisfies the conditions (2) and (3) of the previous definition and the weaker condition*

1 For every $i \in I$, $\overline{p} \cdot \overline{x}_i \leq \overline{p} \cdot \omega_i + \sum_{j \in J} \theta_{i,j} \overline{p} \cdot \overline{y}_j$ and

$$x_i \in P_i(\overline{x}, \overline{y}, \overline{p}) \Longrightarrow \overline{p} \cdot x_i \geq \overline{p} \cdot \overline{x}_i.$$

*The quasiequilibrium $(\overline{x}, \overline{y}, \overline{p})$ is said **nontrivial** if for some $i \in I$ and for some $x_i \in X_i$, $\overline{p} \cdot x_i < \overline{p} \cdot \overline{x}_i$.*

As previously, it follows from Condition 3 that at a quasiequilibrium $(\overline{x}, \overline{y}, \overline{p})$, the equality $\overline{p} \cdot \overline{x}_i = \overline{p} \cdot \omega_i + \sum_{j \in J} \theta_{i,j} \overline{p} \cdot \overline{y}_j$ holds for every $i \in I$.

The importance of nontriviality for quasiequilibrium will be seen later, in Section 2.3.

If the closed convex cone Z is equal to the negative orthant $-\mathbb{R}^{\ell}_{+}$ of $\mathbb{R}^{\ell}$, we have as a special case the concept of **free-disposal equilibrium** (or **free-disposal quasi-equilibrium**). All commodities can be freely disposed of, and Condition (3) of Definitions 2.0.4 and 2.0.5 guarantees that the prices of all commodities are greater or equal to zero and that, on the market for each commodity, supply equals demand or supply exceeds demand with the corresponding equilibrium value equal to zero.

If the convex cone Z is reduced to $\{0\}$, we have the concept of **equilibrium without disposal** (or **quasiequilibrium without disposal**), often simply called **exact equilibrium** or **equilibrium** (resp. **exact quasiequilibrium** or **quasiequilibrium**). Then, since supply equals demand on the market for each commodity, the condition (3) automatically holds.

Investigation of the existence of competitive equilibrium is the first problem in the theory of competitive equilibrium. This problem was solved between 1954 (Arrow–Debreu [21]) and 1962 (Debreu [59]) in the case where consumers' preferences are derived from a complete preorder of each i on his consumption set, and under various sets of other hypothesis. In subsequent papers, several authors have tried to simplify proofs which, as in McKenzie [136], [137] and Debreu [59], are mathematically complex and not very intuitive.

This chapter deals with the existence of competitive equilibrium when consumers have a complete preference preorder on their consumption set. It is that we call *Transitive Equilibrium.* Observe that the hypothesis which defines Transitive Equilibrium requires a strong rationality of consumers' behavior. First, consumers should be able to compare any pair of commodity bundles in their consumption set. Second, the transitivity of preferences implies a strong coherence of their choices which is not of more empirical evidence[2] than the first requirement. Moreover, this hypothesis prevents to consider dependent consumers' preferences, contrary to the game theoretic spirit which inspires the more general framework presented in the introduction.

Chapter 2 is organized as follows.

[2]Think, for example, of multicriteria choices.

In order to allow for an autonomous reading of the chapter, we state and prove in Section 2.1, a generalized variant of the Debreu–Gale–Nikaido lemma. This variant appeared in Chapter 1 as a corollary of the Ky Fan coincidence theorem. The direct proof given here is less demanding from the point of view of the necessary mathematical background. A by-product of the result is an elementary proof of Kakutani's theorem. Based on this lemma, the equilibrium existence proof performed in Sections 2.2 and 2.3 uses several now classical tricks. First, following a procedure devised by Debreu [59], the research of sufficient conditions for the existence of a quasiequilibrium is clearly distinguished from a general study of the conditions under which a quasiequilibrium is actually an equilibrium. However, the formal definition given above is not Debreu's definition but the today standard quasiequilibrium definition, first proposed by Bergstrom [29], well-suited to such a separation of problems.[3]

To prove in Section 2.2 the existence of a so-defined quasiequilibrium, we use several other tricks borrowed from various authors and that we will use again in the next chapters of this book. Bounding the economy so as to apply a fixed point theorem to an artificial compact economy before letting the bounds go to infinity is an as old trick as the first equilibrium existence proofs.[4] Classically, we first study a compactified economy before proving the quasiequilibrium existence in the original economy. When stating sufficient conditions for quasiequilibrium existence, the definition and the role of "augmented preferences" is borrowed from Gale and Mas-Colell (see [94] and [96]). Introduced in papers devoted to nontransitive equilibrium, this definition allows to replace by insatiability the often assumed local insatiability of consumers at any point of their attainable consumption set. On the other hand, as we let the unit-ball of $\mathbb{R}^\ell$ be the domain of admissible prices, problems arise with a lack of continuity at the origin of certain correspondences (here the excess quasidemand correspondence to which we want to apply the variant of the Debreu–Gale–Nikaido lemma stated in Section 2.1). To solve this problem, we borrow from Bergstrom [30] the definition of modified budget sets which coincide with the original ones when prices belong to the unit-sphere.

[3]This definition coincides with the Arrow–Hahn [22] definition of "compensated equilibrium" after a convenient adaptation of this notion to our general framework and if one assumes that consumers are locally no-satiated at equilibrium.

[4]The idea goes back to A. Wald [176].

To pass in Section 2.3 from the quasiequilibrium obtained in Section 2.2 to an equilibrium requires an additional continuity assumption on consumer preferences and, in the absence of strong autarchy assumptions for consumers, an irreducibility assumption on the economy. We attempt in this section to unify various irreducibility approaches found in the literature. As the results obtained in this section do not depend on the particular assumption on consumers' preferences which define Transitive Equilibrium, we state and prove them in the most general framework. In this guise, these results will also be used in the next chapters.

In counterpart, let us stress at the end of this introduction that the excess (quasi)demand approach used in Section 2.2 strongly relies not only on the transitivity but also on the completeness of consumers' preference preorders on their consumption set.[5] Without this assumption, this approach could not be used for proving the existence of nontransitive quasiequilibrium without strengthening the other hypothesis. The interest of the next chapters will be to show that other approaches allow for proving the equilibrium existence without transitivity or completeness of preferences and under even weaker other assumptions.

2.1 A direct proof of a Debreu–Gale–Nikaido lemma

The fixed point lemma which will be used in the next section is the corollary 1.3.5 of the Ky Fan coincidence theorem (Proposition 1.3.2). The proof given here is based on Brouwer's fixed point theorem and the concept of continuous partition of unity weakly subordinate to a finite open covering of a compact subset of $\mathbb{R}^\ell$.

Recall first the continuity definition we use for a correspondence defined on a subset X of $\mathbb{R}^\ell$ and with values in $\mathbb{R}^\ell$.

DEFINITION 2.1.1 *A correspondence $\varphi: X \to \mathbb{R}^\ell$ is called* ***upper hemicontinuous at*** $x^0 \in X$ *if for every $p \in \mathbb{R}^\ell$, the support function*

[5]Even in the absence of externalities, that is if an excess (quasi)demand can be defined as a correspondence depending only on prices, without completeness of the partial preference preorders, the convexity of values of the excess (quasi)demand correspondence is no longer insured by the convexity assumptions made on preferences. Mas-Colell [126] gives the example of an irreflexive, transitive and convex valued strict preference relation defined on the positive cone of $\mathbb{R}^2$, which is not the asymmetric part of a complete preorder, such that for some price and some wealth w the set $\varphi(p, w)$ is not convex.

$x \mapsto \sup_{y \in \varphi(x)} p \cdot y$ (equivalently $x \mapsto \inf_{y \in \varphi(x)} p \cdot y$) is upper semicontinuous (resp. lower semicontinuous) at x^0.

The correspondence φ is called ***upper hemicontinuous on*** *X (or simply* ***upper hemicontinuous****) if φ is upper hemicontinuous at every point of X.*

It is readily seen that the upper semicontinuity of a correspondence, as usually defined (see Definition 1.1.1), implies its upper hemicontinuity. If φ has closed convex values in a fixed compact subset of $\mathbb{R}^\ell$, the upper hemicontinuity of φ at every point of X coincides with its upper semicontinuity on X. It is no longer true if $\varphi(X)$ is not a relatively compact subset of $\mathbb{R}^\ell$.

Maybe the most convenient version of the Debreu–Gale–Nikaido lemma for applications to transitive equilibrium is the following one, where correspondence ζ can be thought of as an excess demand correspondence and condition 1.3 as a weakening of Walras's law.

LEMMA 2.1.1 *Let P be a closed convex cone (with vertex 0) of $\mathbb{R}^\ell$, and $\overline{B} := \{x \in \mathbb{R}^\ell : \|x\| \leq 1\}$, $S := \{x \in \mathbb{R}^\ell : \|x\| = 1\}$ be respectively the (closed) unit ball and the unit-sphere of $\mathbb{R}^\ell$. Let $\zeta : \overline{B} \cap P \to \mathbb{R}^\ell$ be a nonempty compact convex valued, upper hemicontinuous correspondence. Assume that*

$$\forall p \in S \cap P, \ \exists z \in \zeta(p), \ p \cdot z \leq 0. \tag{2.2}$$

Then there exists $\overline{p} \in \overline{B} \cap P$ such that $\zeta(\overline{p}) \cap P^0 \neq \emptyset$, where

$$P^0 = \{z \in \mathbb{R}^\ell : z \cdot q \leq 0, \ \forall q \in P\}$$

is the polar cone of P.

Proof. Assume by contraposition that the conclusion of the lemma is false. Then for any $p \in \overline{B} \cap P$, the closed convex cone P^0 and the nonempty compact convex set $\zeta(p)$ can be strictly separated by an hyperplane. In other words, for any p in $\overline{B} \cap P$, there is some q in $\overline{B}$ and some real number α such that

$$\sup_{z \in P^0} q \cdot z \leq \alpha < \inf_{z \in \zeta(p)} q \cdot z \,.$$

It is easy to see that $\alpha \geq 0$ and consequently that $q \in P^{00} = P$. On the other hand, since the correspondence ζ is upper hemicontinuous, the subsets

$$V(q) = \left\{p \in \overline{B} \cap P : \inf_{z \in \zeta(p)} q \cdot z > 0\right\}$$

are open in $\overline{B} \cap P$ and the family $(V(q))_{q \in \overline{B} \cap P}$ is an open covering of $\overline{B} \cap P$. As $\overline{B} \cap P$ is compact, there exist $q^i \in \overline{B} \cap P$, $i = 1, \ldots, n$, such that

$$(\overline{B} \cap P) \subset \bigcup_{i=1}^{n} V(q^i) .$$

Let $(f_i)_{i=1}^n$ be a continuous partition of unity weakly subordinate to this finite open covering, that is a family of continuous real functions defined on $\overline{B} \cap P$, such that $0 \leq f_i(p) \leq 1$ for all i and all p in $\overline{B} \cap P$, $f_i(p) = 0$ outside $V(q^i)$ and $\sum_{i=1}^n f_i(p) = 1$ for all p in $\overline{B} \cap P$.

Define the functions $f \colon \overline{B} \cap P \to \overline{B} \cap P$ and $g \colon \overline{B} \cap P \to S \cap P$ by

$$f(p) = \sum_{i=1}^{n} f_i(p) q^i \quad \text{and} \quad g(p) = \frac{f(p)}{\|f(p)\|} .$$

If $z \in \zeta(p)$,

$$f_i(p) \neq 0 \Longrightarrow p \in V(q^i) \Longrightarrow q^i \cdot z > 0 ,$$

which implies

$$f(p) \cdot z > 0 .$$

Thus $f(p) \neq 0$ for all p in $\overline{B} \cap P$ and g is a continuous function from $\overline{B} \cap P$ into $(S \cap P) \subset (\overline{B} \cap P)$.

According to Brouwer's theorem, there exists $\overline{p} \in \overline{B} \cap P$ such that $\overline{p} = g(\overline{p}) \in S \cap P$. Then we have

$$f_i(\overline{p}) \neq 0 \Longrightarrow \inf_{z \in \zeta(\overline{p})} q^i \cdot z > 0 ,$$

and consequently

$$\inf_{z \in \zeta(\overline{p})} \overline{p} \cdot z > 0 .$$

On the other hand, in view of the assumption made on ζ in the lemma and since $\overline{p} \in S \cap P$, there exists $z \in \zeta(\overline{p})$ such that $\overline{p} \cdot z \leq 0$.

We thus obtain a contradiction, proving the lemma. □

It suffices to suppose first that P is the positive orthant $\mathbb{R}^\ell_+$ of $\mathbb{R}^\ell$, and second that P is the whole space for obtaining the two following corollaries:

COROLLARY 2.1.1 *Let $\zeta : \overline{B} \cap \mathbb{R}^\ell_+ \to \mathbb{R}^\ell$ be a nonempty compact convex valued, upper hemicontinuous correspondence. Assume that*

$$\forall p \in S \cap \mathbb{R}^\ell_+, \ \exists z \in \zeta(p), \ p \cdot z \leq 0. \qquad (2.3)$$

Then there exists $\overline{p} \in \overline{B} \cap \mathbb{R}^\ell_+$ *such that* $\zeta(\overline{p}) \cap -\mathbb{R}^\ell_+ \neq \emptyset$.

COROLLARY 2.1.2 (CORNET) *Let* $\zeta : \overline{B} \to \mathbb{R}^\ell$ *be a nonempty closed convex valued, upper hemicontinuous correspondence. Assume that*

$$\forall p \in S,\ \exists z \in \zeta(p),\ p \cdot z \leq 0. \tag{2.4}$$

Then there exists $\overline{p} \in \overline{B}$ *such that* $0 \in \zeta(\overline{p})$.

Note that in Corollary 2.1.2 the correspondence ζ is not supposed to be compact valued but only closed valued. A glance at the proof of the lemma shows that this weakening is possible: if $0 \notin \zeta(p)$, then the subsets $\{0\}$ and $\zeta(p)$ can be strictly separated by an hyperplane.

Cornet [54] proves Corollary 2.1.2 as a consequence of a minimax theorem of Ky Fan [75]. As already recalled in the preamble of this chapter, the proof given in Chapter 1 (see Corollary 1.3.2) was based on the Ky Fan Correspondence Theorem. A weaker version of the same result is implicitly used and proved by Bergstrom [30] from Kakutani's theorem.

Regarding the assumptions, Lemma 2.1.1 and its two corollaries, with the weakened Walras Law condition they state, improve upon the corresponding upper semicontinuous versions (which are in the literature proved from Kakutani's theorem). With regards to the statement of Debreu [57], Lemma 2.1.1 has the advantage of not ruling out the case where P is a linear subspace of $\mathbb{R}^\ell$. Note however that the point $\overline{p}$, of which Lemma 2.1.1 and its corollaries establish the existence, is not proved to be $\neq 0$, as it is shown in the standard statements of Lemma 2.1.1 ([57]), [104]) and Corollary 2.1.1 ([58, 5.6], [147, Theorem 16.6]). This weakening of the conclusion is likely to be the counterpart of the advantage we see in the fact that the validity of Lemma 2.1.1 is not limited to the case where the cone P is not a linear subspace.

Uzawa [175] had already noticed that Kakutani's theorem can be deduced from the Debreu–Gale–Nikaido lemma. We next show that Corollary 2.1.2 also implies Kakutani's theorem.

COROLLARY 2.1.3 (KAKUTANI) *Let* X *be a nonempty compact convex subset of* $\mathbb{R}^\ell$ *and* $\varphi\colon X \to X$ *an upper semicontinuous and nonempty closed convex valued correspondence. Then there exist* $\overline{x} \in X$ *such that* $\overline{x} \in \varphi(\overline{x})$.

Proof. It is sufficient to prove the theorem when X is the unit-ball. Let $I_{\overline{B}}\colon \overline{B} \to \overline{B}$ be the identity correspondence $I_{\overline{B}}(x) = \{x\}$ and $\zeta\colon \overline{B} \to \mathbb{R}^\ell$

the correspondence $\zeta = \varphi - I_{\overline{B}}$. Clearly, ζ is upper semicontinuous, therefore upper hemicontinuous, and nonempty closed convex valued. On the other hand, if $p \in S$ and $z = y - p$ with $y \in \varphi(p)$, then

$$p \cdot z = p \cdot y - \|p\|^2 = p \cdot y - 1 \leq 0 \,.$$

Applying Corollary 2.1.2, there exists $\overline{p} \in \overline{B}$ such that $0 \in \varphi(\overline{p})$, that is $\overline{p} \in \varphi(\overline{p})$. □

2.2 Existence of transitive quasiequilibrium

We assume in this section that

$$\mathcal{E} = \left(\langle \mathbb{R}^\ell, \mathbb{R}^\ell \rangle, (X_i, \succsim_i, \omega_i)_{i \in I}, (Y_j)_{j \in J}, (\theta_{i,j})_{\substack{i \in I \\ j \in J}}, Z \right)$$

where for each i, $\succsim_i$ denotes a complete preorder representing the preferences of the ith consumer over the commodity bundles of his consumption set X_i. With the notations introduced in the preamble of the chapter, for $(x, y, p) \in \prod_{i \in I} X_i \times \prod_{j \in J} Y_j \times S$, we have now:

$$P_i(x, y, p) = \{x'_i \in X_i \,:\, x'_i \succ_i x_i\},$$

where the strict preference $\succ_i$ is the asymmetrical part of the complete preorder $\succsim_i$. For convenience, we will call $P_i(x_i)$ this set which does not depend on $((x_k)_{k \neq i}, (y_j)_{j \in J}, p)$, considering the preference correspondence P_i as a correspondence defined on X_i and with values in X_i .

Before stating the assumptions considered on $\mathcal{E}$, let us first introduce some notation:

- $\omega = \sum\limits_{i \in I} \omega_i$ is the total initial endowment;
- $Y = \sum\limits_{j \in J} Y_j$ is the total production set;
- $\mathcal{A}(\mathcal{E}) = \left\{ (x, y) \in \prod\limits_{i \in I} X_i \times \prod\limits_{j \in J} Y_j \,:\, \sum\limits_{i \in I} x_i - \sum\limits_{j \in J} y_j - \omega \in Z \right\}$ is the set of all attainable allocations;
- $\widehat{X} = \left\{ x \in \prod\limits_{i \in I} X_i \,:\, \exists y \in Y,\ \sum\limits_{i \in I} x_i - y - \omega \in Z \right\}$ is the set of all attainable consumption allocations;
- $\widehat{Y} = \left\{ y \in Y \,:\, \exists x \in \prod\limits_{i \in I} X_i,\ \sum\limits_{i \in I} x_i - y - \omega \in Z \right\}$ is the attainable total production set;

- $\widehat{X}_i = \left\{x_i \in X_i : \exists((x_{i'})_{i' \neq i}, y) \in \prod_{i' \neq i} X_{i'} \times Y,\ x_i + \sum_{i' \neq i} x_{i'} - y - \omega \in Z\right\}$ is the attainable set of the ith consumer;
- $\widehat{Y}_j = \left\{y_j \in Y_j : \exists(x, (y_{j'})_{j' \neq j}) \in \prod_{i \in I} X_i \times \prod_{j' \neq j} Y_{j'},\ \sum_{i \in I} x_i - y_j - \sum_{j' \neq j} y_{j'} - \omega \in Z\right\}$ is the attainable set of the jth producer.

We will prove the existence of a quasiequilibrium under the following assumptions on $\mathcal{E}$:

1 For each $i \in I$,

(a) X_i is convex and closed,

(b) For each $x_i \in X_i$, $(P_i)^{-1}(x_i)$ is open in X_i,

(c) For each $x_i \in X_i$, $P_i(x_i)$ is convex,

(d) $\omega_i \in X_i - \sum_{j \in J} \theta_{i,j} Y_j - Z$ and $\widehat{X}_i$ is compact,

(e) For each $x_i \in \widehat{X}_i$, one has $P_i(x_i) \neq \emptyset$;

2 For each $j \in J$, Y_j is convex and closed;

3 (a) Z is a closed convex cone contained in the negative orthant $-\mathbb{R}^{\ell}_{+}$,

(b) $Y + Z$ is closed.

Closedness and convexity are classical assumptions on consumption sets as well as on production sets. Note that they imply in particular that commodities are perfectly divisible. Assumption 1 (b) is a continuity assumption on preferences. Under this hypothesis, as proved by Rader [150], Debreu [60], each complete preorder of preferences $\succsim_i$ on X_i can be represented by an upper semicontinuous utility function. Convexity of strictly preferred sets is a strong assumption which rules out several phenomena as non-risk aversion. The first part of Assumption 1 (d) is an autarky assumption: using his own shares in the productive system and after disposal of an eventual surplus, consumer i can survive without participating in any exchange.[6] Made for every i, this assumption insures that $\mathcal{E}$ has attainable allocations. Proposition 2.2.4 will give

[6]From an empirical point of view, such an assumption is obviously of a questionable plausibility. It may be strengthened to the survival assumption: $\omega_i \in X_i - Z$, joint to the assumption that $0 \in Y_j$ for each $j \in J$, that is, that inaction is possible for every producer.

assumptions on the primitive characteristics of the economy which guarantee the closedness of $Y+Z$ (Assumption 3 (b)) and the compactness of all individual attainable consumption sets, as assumed in Assumptions 1 (d). Assumption 1 (e) assumes, for every i, the insatiability of the ith consumer at any point of his attainable consumption set.

2.2.1 Compactifying the economy

It follows from the completeness of preference preorders, the compactness of each $\widehat{X}_i$ and Assumption 1 (b) that each $\widehat{X}_i$ has a most preferred element $b_i \in \widehat{X}_i$, $b_i \succsim_i x_i$, $\forall x_i \in \widehat{X}_i$. Consequently, for each i, we can fix $a_i \in X_i$ such that $a_i \succ_i x_i$, $\forall x_i \in \widehat{X}_i$. Let us fix also $\underline{x}_i \in X_i$, and $\underline{y}_{i,j} \in Y_j$ such that, as in Assumption 1 (d), $\underline{x}_i - \omega_i - \sum_{j \in J} \theta_{i,j} \underline{y}_{i,j} \in Z$ for every $i \in I$.

Let $(r^k)_{k \in \mathbb{N}}$ be an increasing sequence of positive real numbers converging to $+\infty$ and such that the closed ball $\overline{B}(0, r^k)$ with center 0 and radius r^k contains all $\widehat{X}_i$, $\underline{x}_i$, $\underline{y}_{i,j}$ and a_i. For each k, let us define $X_i^k = X_i \cap \overline{B}(0, r^k)$, $Y_j^k = Y_j \cap \overline{B}(0, r^k)$ and let $\succsim_i^k$ denote the complete preorder induced on X_i^k by $\succsim_i$. We now consider the "compact" economy

$$\mathcal{E}^k = \Big(\langle \mathbb{R}^\ell, \mathbb{R}^\ell \rangle, (X_i^k, \succsim_i^k, \omega_i)_{i \in I}, (Y_j^k)_{j \in J}, (\theta_{i,j})_{\substack{i \in I \\ j \in J}}, Z \Big)$$

and observe that $\mathcal{E}^k$ satisfies the same assumptions as $\mathcal{E}$.

On the other hand, let us borrow from Gale and Mas-Colell ([94], [96]) the definition of "augmented" preference correspondences:

$$\widehat{P}_i^k(x_i) = \{ x_i' \in X_i^k \, : \, x_i' = x_i + \lambda(x_i'' - x_i), 0 < \lambda \leq 1, \; x_i'' \in P_i^k(x_i) \}.$$

Recall that $\overline{B} = \{x \in \mathbb{R}^\ell : \|x\| \leq 1\}$ and $S = \{x \in \mathbb{R}^\ell : \|x\| = 1\}$ denote respectively the (closed) unit ball and the unit-sphere of $\mathbb{R}^\ell$. Since each Y_j^k is compact, we can define for every $p \in \overline{B}$ the functions and the correspondences:

$$\begin{aligned}
&\pi_j^k(p) = \sup p \cdot Y_j^k = \sup\{p \cdot y_j \, : \, y_j \in Y_j^k\}, \\
&\psi_j^k(p) = \{y_j \in Y_j^k \, : \, p \cdot y_j = \pi_j^k(p)\}, \\
&w_i^k(p) = p \cdot \omega_i + \textstyle\sum_{j \in J} \theta_{i,j} \pi_j^k(p), \\
&\gamma_i^k(p) = \{x_i \in X_i^k \, : \, p \cdot x_i \leq w_i^k(p)\}, \\
&\delta_i^k(p) = \{x_i \in X_i^k \, : \, p \cdot x_i < w_i^k(p)\}, \\
&\xi_i^k(p) = \{x_i \in X_i^k \, : \, x_i \in \gamma_i^k(p) \text{ and } \widehat{P}_i^k(x_i) \cap \gamma_i^k(p) = \emptyset\},
\end{aligned}$$

$\varphi_i^k(p) = \{x_i \in X_i^k : x_i \in \gamma_i^k(p) \text{ and } \widehat{P}_i^k(x_i) \cap \delta_i^k(p) = \emptyset\}$.

It easily follows from the Maximum theorem[7] that for each $j \in J$, the function $\pi_j^k : \overline{B} \to \mathbb{R}$ is continuous and that the correspondence $\psi_j^k : \overline{B} \to Y_j^k$ has a closed graph with nonempty convex values (equivalently, is upper semicontinuous with nonempty, closed, convex values). Each ψ_j^k can be thought of as the *supply correspondence* of producer j in the economy $\mathcal{E}^k$. For each $i \in I$, the function $w_i^k : \overline{B} \to \mathbb{R}$ is continuous. The correspondence $\gamma_i^k : \overline{B} \to X_i^k$ is the *budget correspondence* of i in $\mathcal{E}^k$ when the wealth of consumer i is defined by w_i^k. The correspondence ξ_i^k can be thought of as the *demand correspondence* of i in $\mathcal{E}^k$ after substituting correspondence $\widehat{P}_i^k$ to correspondence P_i^k. We will say that $\varphi_i^k : \overline{B} \to X_i^k$ is the *quasidemand correspondence* of i in $\mathcal{E}^k$ after substituting correspondence $\widehat{P}_i^k$ to correspondence P_i^k.

We now state and prove a first result in the compact economy $\mathcal{E}^k$.

PROPOSITION 2.2.1 *Under the assumptions 1, 2, 3 stated above on the economy $\mathcal{E}$, for any $k \geq 1$ there exist $(\overline{x}^k, \overline{y}^k, \overline{p}^k) \in \prod_{i \in I} X_i^k \times \prod_{j \in J} Y_j^k \times (S \cap Z^0)$ such that*

1 $\sum_{i \in I} \overline{x}_i^k - \sum_{j \in J} \overline{y}_j^k - \omega \in Z$;

2 For each $i \in I$, $\overline{p}^k \cdot \overline{x}_i^k = \overline{p}^k \cdot \omega_i + \sum_{j \in J} \theta_{i,j} \overline{p}^k \cdot \overline{y}_j^k$ *and* $\widehat{P}_i^k(\overline{x}_i^k) \cap \delta_i^k(\overline{p}^k) = \emptyset$;

3 For each $j \in J$, *for every* $y_j \in Y_j^k$, $\overline{p}^k \cdot y_j \leq \overline{p}^k \cdot \overline{y}_j^k$.

4 $\overline{p}^k \cdot (\sum_{i \in I} \overline{x}_i^k - \sum_{j \in J} \overline{y}_j^k - \omega) = 0$.

Equivalently, the economy $\mathcal{E}^k$ has a quasiequilibrium.

Proof. Following an idea borrowed from Bergstrom [30], let us define on $\overline{B}$ the correspondences $\gamma_i'^k$, $\delta_i'^k$, $\xi_i'^k$, $\varphi_i'^k$, which coincide with the above defined correspondences γ_i^k, δ_i^k, ξ_i^k, φ_i^k when $p \in S$:

$\gamma_i'^k(p) = \{x_i \in X_i^k : p \cdot x_i \leq w_i^k(p) + \frac{1 - \|p\|}{m}\}$,

[7] See Appendix.

$\delta_i'^k(p) = \{x_i \in X_i^k : p \cdot x_i < w_i^k(p) + \frac{1-\|p\|}{m}\}$,
$\xi_i'^k(p) = \{x_i \in X_i^k : x_i \in \gamma_i^k(p) \text{ and } \widehat{P}_i^k(x_i) \cap \gamma_i'^k(p) = \emptyset\}$,
$\varphi_i'^k(p) = \{x_i \in X_i^k : x_i \in \gamma_i^k(p) \text{ and } \widehat{P}_i^k(x_i) \cap \delta_i'^k(p) = \emptyset\}$.

In order to apply Lemma 2.1.1 to the correspondence $\zeta^k: \overline{B} \to \mathbb{R}^\ell$,

$$p \mapsto \zeta^k(p) := \sum_{i \in I} \varphi_i'^k(p) - \sum_{j \in J} \psi_j^k(p) - \{\omega\},$$

we next prove that each $\varphi_i'^k$ is upper semicontinuous with nonempty closed convex values for all $p \in \overline{B} \cap Z^0$ (equivalently that $\varphi_i'^k: (\overline{B} \cap Y^0) \to X_i^k$ has a closed graph with nonempty convex values).

Closedness of $\varphi_i'^k$. Let $(p^\nu)_{\nu \in \mathbb{N}}$ and $(x_i^\nu)_{\nu \in \mathbb{N}}$ be two sequences in $\overline{B} \cap Z^0$ and X_i^k respectively, converging to $p \in \overline{B} \cap Z^0$ and $x_i \in X_i^k$ and such that $x_i^\nu \in \varphi_i'^k(p^\nu)$ for all $\nu \in \mathbb{N}$. Clearly, $\gamma_i'^k$ has a closed graph in $\overline{B} \times X_i^k$ and it follows from $x_i^\nu \in \gamma_i'^k(p^\nu)$ for each $\nu \in \mathbb{N}$ that $x_i \in \gamma_i'^k(p)$.

Let us divide the argument in two cases:

1 If $p \cdot x_i < w_i^k(p) + \frac{1-\|p\|}{m}$ and if there exists $y_i^k \in P_i^k(x_i)$, let ν' and, according to Assumption 1 (b), ν'' be such that

$$\nu > \nu' \Longrightarrow p \cdot x_i^\nu < w_i^k(p^\nu) + \frac{1 - \|p^\nu\|}{m},$$

$$\nu > \nu'' \Longrightarrow x_i^\nu \in (P_i^k)^{-1}(y_i^k), \text{ that is to say } y_i^k \in P_i^k(x_i^\nu).$$

From the definition of $\widehat{P}_i^k$ it follows that

$$\nu > \max\{\nu', \nu''\} \Longrightarrow \delta_i'^k(p^\nu) \cap \widehat{P}_i^k(x_i^\nu) \neq \emptyset,$$

which contradicts the assumption made on x_i^ν.

2 If $p \cdot x_i = w_i^k(p) + \frac{1-\|p\|}{m}$ and if there exists $y_i^k \in \widehat{P}_i^k(x_i) \cap \delta_i'^k(p)$, then from the definition of $\widehat{P}_i^k$, it follows that there exists $y_i^k \in P_i^k(x_i) \cap \delta_i'^k(p)$.

Let ν' and, according to Assumption 1 (b), ν'' be such that

$$\nu > \nu' \Longrightarrow y_i^k \in \delta_i'^k(p^\nu),$$

$$\nu > \nu'' \Longrightarrow y_i^k \in P_i^k(x_i^\nu).$$

We have

$$\nu > \max\{\nu', \nu''\} \Longrightarrow y_i^k \in (P_i^k(x_i^\nu) \cap \delta_i'^k(p^\nu)) \subset (\widehat{P}_i^k(x_i^\nu) \cap \delta_i'^k(p^\nu)),$$

which contradicts the assumption made on x_i^ν.

We thus have:

$$\begin{array}{ll} P_i^k(x_i) = \emptyset & \text{if } \ p \cdot x_i < w_i^k(p) + \frac{1-\|p\|}{m} \\ \widehat{P}_i^k(x_i) \cap \delta_i'^k(p) = \emptyset & \text{if } \ p \cdot x_i = w_i^k(p) + \frac{1-\|p\|}{m} \end{array}.$$

In both cases, we have proved that $x_i \in \varphi_i'^k(p)$.

Nonemptiness of each $\varphi_i'^k(p)$ for all $p \in (\overline{B} \cap Z^0)$. Since $\xi_i'^k(p) \subset \varphi_i'^k(p)$, it is sufficient to prove that for each $p \in \overline{B} \cap Z^0$, $\xi_i'^k(p)$ is nonempty. To see this, let $p \in \overline{B} \cap Z^0$. It follows from the first part of Assumption 1 (d) on $\mathcal{E}$ that $\gamma_i'^k(p) \neq \emptyset$. On the other hand, it follows from the compactness of this set, Assumption 1 (b) and the completeness of the preorder $\succsim_i$ that there exists $x_i \in \gamma_i'^k(p)$ such that $\gamma_i'^k(p) \cap P_i^k(x_i) = \emptyset$.

If $P_i^k(x_i) = \emptyset$, it obviously follows that $\gamma_i'^k(p) \cap \widehat{P}_i^k(x_i) = \emptyset$ and $x_i \in \xi_i'^k(p)$.

If $P_i^k(x_i) \neq \emptyset$, note first that $p \neq 0$ follows from $\gamma_i'^k(p) \cap P_i^k(x_i) = \emptyset$. Consequently, let $x_i' \in P_i^k(x_i)$ and x_i'' be the point where the line segment joining x_i and x_i' intersects the hyperplane of equation $p \cdot x_i = w_i^k(p) + \frac{1-\|p\|}{m}$. By the definition of x_i and since $\succsim_i$ is a complete preorder, $x_i \succsim_i x_i''$. By transitivity, one also has $x_i' \succ_i x_i''$. In view of Assumption 1. (c), $x_i \succ_i x_i''$ is impossible, thus $P_i^k(x_i) = P_i^k(x_i'')$ and $\gamma_i'^k(p) \cap P_i^k(x_i'') = \emptyset$. As $p \cdot x_i'' = w_i^k(p) + \frac{1-\|p\|}{m}$, $\gamma_i'^k(p) \cap \widehat{P}_i^k(x_i'') = \emptyset$ and $x_i'' \in \xi_i'^k(p)$.

Convexity of each $\varphi_i'^k(p)$ for all $p \in (\overline{B} \cap Z^0)$. Let $p \in (\overline{B} \cap Z^0)$, x_i and x_i' in $\varphi_i'^k(p)$ and let x_i'' be on the line segment joining x_i and x_i'.

If either x_i or x_i' belongs to $\delta_i'^k(p)$, it follows from the definition of $\varphi_i'^k$ and the completeness of preorder $\succsim_i$ that x_i and x_i' are equivalent for i. It then follows from the definition of $\varphi_i'^k$ and Assumption 1 (c) that x_i'' is equivalent for i to x_i and x_i', thus that x_i'' belongs to $\varphi_i'^k(p)$.

If x_i and x_i' belong to the hyperplane of equation $p \cdot x_i = w_i^k(p) + \frac{1-\|p\|}{m}$, the same is true for x_i'' and

$$\widehat{P}_i^k(x_i'') \cap \delta_i'^k(p) \neq \emptyset \Longrightarrow P_i^k(x_i'') \cap \delta_i'^k(p) \neq \emptyset\,.$$

In order to prove that x_i'' belongs to $\varphi_i'^k(p)$, assume by contraposition that there is some z_i in $P_i^k(x_i'') \cap \delta_i'^k(p)$. From the transitivity and completeness of preorder and the definition of $\varphi_i'^k$, it is easily seen that x_i and x_i' belong to $P_i^k(x_i'')$, in contradiction with Assumption 1 (c) and the irreflexivity of the strict preference.

Applying Lemma 2.1.1. From what precedes, it follows that the correspondence $\zeta^k\colon (\overline{B} \cap Z^0) \to \mathbb{R}^\ell$ is upper semicontinuous (thus upper hemicontinuous) with nonempty compact convex values. In order to apply Lemma 2.1.1, it remains to verify the boundary condition (1.3). To do it, let $p \in (S \cap Z^0)$ and $z \in \zeta^k(p) = \sum_{i\in I} \varphi_i'^k(p) - \sum_{j\in J} \psi_j^k(p) - \{\omega\}$. From $\|p\| = 1$, we deduce $z \in \sum_{i\in I} \varphi_i^k(p) - \sum_{j\in J} \psi_j^k(p) - \{\omega\}$, that is $z = \sum_{i\in I} x_i - \sum_{j\in J} y_j - \{\omega\}$ with each $x_i \in \varphi_i^k(p)$ and each $y_j \in \psi_j^k(p)$. In particular, each $x_i \in \gamma_i^k(p)$ and for each i,

$$p \cdot x_i \leq p \cdot \omega_i + \sum_{j\in J} \theta_{i,j} p \cdot y_j.$$

Summing on $i \in I$ and using the relations $\sum_{i\in I} \theta_{i,j} = 1$, $\forall j \in J$, one gets as required $p \cdot z \leq 0$. Note that here the boundary condition is satisfied for all $z \in \zeta^k(p)$.

In view of the conclusion of Lemma 2.1.1, let $\overline{p}^k \in \overline{B} \cap Z^0$ and $\overline{z}^k \in \zeta(\overline{p}^k) \cap Z^{00} = \zeta(\overline{p}^k) \cap Z$. Thus there exist for each $i \in I$, $\overline{x}_i^k \in \varphi_i'^k(\overline{p}^k)$, and for each $j \in J$, $\overline{y}_j^k \in \psi_j^k(\overline{p}^k)$ with $\overline{z}^k = \sum_{i\in I} \overline{x}_i^k - \sum_{j\in J} \overline{y}_i^k - \omega \in Z$. In other words, the allocation $((\overline{x}_i^k)_{i\in I}, (\overline{y}_j^k)_{j\in J})$ is attainable. Using the definition of φ_i^k, the definition of $\widehat{P}_i^k$, and Assumption 1 (e), one easily sees that for each i,

$$\overline{p}^k \cdot \overline{x}_i^k = w_i^k(\overline{p}^k) + \frac{1 - \|\overline{p}^k\|}{m} = \overline{p}^k \cdot \omega_i + \sum_{j\in J} \theta_{i,j} \overline{p}^k \cdot \overline{y}_j^k + \frac{1 - \|\overline{p}^k\|}{m}.$$

Summing on i, one gets $\overline{p}^k \cdot \overline{z}^k = 1 - \|\overline{p}^k\| \geq 0$. Together with $\overline{p}^k \in Z^0$ this implies $\|p\| = 1$, $p \in S$ and $\gamma_i'^k(\overline{p}^k) = \gamma_i^k(\overline{p}^k)$, $\delta_i'^k(\overline{p}^k) = \delta_i^k(\overline{p}^k)$, $\varphi_i'^k(\overline{p}^k) = \varphi_i^k(\overline{p}^k)$. Summing once more on i the relations $\overline{p}^k \cdot \overline{x}_i^k = \overline{p}^k \cdot \omega_i + \sum_{j\in J} \theta_{i,j} \overline{p}^k \cdot \overline{y}_j^k$, one has finally $\overline{p}^k \cdot \overline{z}^k = 0$.

Moreover, for each i, as $\overline{p}^k \cdot \overline{x}_i^k = \overline{p}^k \cdot \omega_i + \sum_{j\in J} \theta_{i,j} \overline{p}^k \cdot \overline{y}_j^k$,

$$\widehat{P}_i^k(\overline{x}_i^k) \cap \delta_i^k(\overline{p}^k) = \emptyset \iff P_i^k(\overline{x}_i^k) \cap \delta_i^k(\overline{p}^k) = \emptyset$$

which ends the proof of Proposition 2.2.1. □

2.2.2 Existence of quasiequilibrium in $\mathcal{E}$

We now let k tend to infinity and pass to limit. Before stating the quasiequilibrium result, as in the compact economy, we define for each i the augmented preference:

$$\widehat{P}_i(x_i) = \{x'_i \in X_i : x'_i = x_i + \lambda(x''_i - x_i), 0 < \lambda \leq 1, \ x''_i \in P_i(x_i)\}.$$

PROPOSITION 2.2.2 *Under the assumptions 1, 2, 3 stated above on the economy $\mathcal{E}$, there exist $(\overline{x}, \overline{y}, \overline{p}) \in \prod_{i\in I} X_i \times \prod_{j\in J} Y_j \times (S \cap Z^0)$ such that*

1 $\sum_{i\in I} \overline{x}_i - \sum_{j\in J} \overline{y}_j - \omega \in Z$;

2 For each $i \in I$, $\overline{p} \cdot \overline{x}_i = \overline{p} \cdot \omega_i + \sum_{j\in J} \theta_{i,j} \overline{p} \cdot \overline{y}_j$ *and* $x_i \in \widehat{P}_i(\overline{x}_i) \Longrightarrow \overline{p} \cdot x_i \geq \overline{p} \cdot \overline{x}_i$;

3 For each $j \in J$, *for every* $y_j \in Y_j$, $\overline{p} \cdot y_j \leq \overline{p} \cdot \overline{y}_j$.

4 $\overline{p} \cdot (\sum_{i\in I} \overline{x}_i - \sum_{j\in J} \overline{y}_j - \omega) = 0$.

Equivalently, the economy $\mathcal{E}$ has a quasiequilibrium.

Proof. Let for each $k \geq 1$, $(\overline{x}^k, \overline{y}^k, \overline{p}^k)$ be as in Proposition 2.2.1. In view of Assumptions 1 (d), let $(\overline{x}, \overline{p})$ be the limit of a convergent subsequence of $(\overline{x}^k, \overline{p}^k)_{k\in\mathbb{N}}$. In view of Assumption 3 (b), there exists $\overline{y} \in \prod_{j\in J} Y_j$ such that

$$\sum_{i\in I} \overline{x}_i - \sum_{j\in J} \overline{y}_j - \omega \in Z. \tag{2.5}$$

Fix now for each $j \in J$, $y_j \in Y_j$, and for each $i \in I$, $x_i \in \widehat{P}_i(\overline{x}_i)$, $z_i \in P_i(\overline{x}_i)$ such that $x_i = \overline{x}_i + \lambda_i(z_i - \overline{x}_i)$ for some $\lambda_i : 0 < \lambda_i \leq 1$. Applying Assumption 1 (b) and the construction of $\mathcal{E}^k$, there exists k' such that $k > k' \Longrightarrow z_i \in X_i^k \cap P_i(\overline{x}_i^k)$, for all $i \in I$ and $y_j \in Y_j^k$, $\overline{y}_j \in Y_j^k$, for all $j \in J$. Defining for each i, $x_i^k = \overline{x}_i^k + \lambda_i(z_i - \overline{x}_i^k)$ we have $x_i^k \in \widehat{P}_i^k(\overline{x}_i^k)$, thus $\overline{p}^k \cdot x_i^k \geq \overline{p}^k \cdot \omega_i + \sum_{j\in J} \theta_{i,j} \overline{p}^k \cdot \overline{y}_j^k$, from which we deduce

$$\overline{p}^k \cdot x_i^k \geq \overline{p}^k \cdot \omega_i + \sum_{j\in J} \theta_{i,j} \overline{p}^k \cdot \overline{y}_j, \tag{2.6}$$

$$\overline{p}^k \cdot x_i^k \geq \overline{p}^k \cdot \omega_i + \sum_{j\in J} \theta_{i,j} \overline{p}^k \cdot y_j. \tag{2.7}$$

Passing to limit in (2.6) and in (2.7), we get

$$\overline{p} \cdot x_i \geq \overline{p} \cdot \omega_i + \sum_{j \in J} \theta_{i,j} \overline{p} \cdot \overline{y}_j, \tag{2.8}$$

$$\overline{p} \cdot x_i \geq \overline{p} \cdot \omega_i + \sum_{j \in J} \theta_{i,j} \overline{p} \cdot y_j. \tag{2.9}$$

Observe now that inequalities (2.8) and (2.9) hold for every $i \in I$ and $j \in J$, whatever be $x_i \in \widehat{P}_i(\overline{x}_i)$ and $y_j \in Y_j$. Using (2.5) and Assumption 1 (e), we remark that for each i and every $y = (y_j)_{j \in J} \in \prod_{j \in J} Y_j$, $\overline{x}_i \in \operatorname{cl} \widehat{P}_i(\overline{x}_i)$, thus

$$\overline{p} \cdot \overline{x}_i \geq \overline{p} \cdot \omega_i + \sum_{j \in J} \theta_{i,j} \overline{p} \cdot \overline{y}_j, \tag{2.10}$$

$$\overline{p} \cdot \overline{x}_i \geq \overline{p} \cdot \omega_i + \sum_{j \in J} \theta_{i,j} \overline{p} \cdot y_j. \tag{2.11}$$

Summing over i the relations (2.10) and using again (2.5), we get

$$\overline{p} \cdot (\sum_{i \in I} \overline{x}_i - \sum_{j \in J} \overline{y}_j - \omega) = 0 \tag{2.12}$$

and for every i,

$$\overline{p} \cdot \overline{x}_i = \overline{p} \cdot \omega_i + \sum_{j \in J} \theta_{i,j} \overline{p} \cdot \overline{y}_j. \tag{2.13}$$

Finally, let $j \in J$ and i such that $\theta_{i,j} > 0$. Applying (2.11) with $y' = (y'_j)_{j \in J} \in \prod_{j \in J} Y_j$ defined by $y'_{j'} = \begin{cases} y_j & \text{if } j' = j \\ \overline{y}_{j'} & \text{if } j' \neq j \end{cases}$, it readily follows from (2.13) that

$$\overline{p} \cdot y_j \leq \overline{p} \cdot \overline{y}_j \quad \forall y_j \in Y_j. \tag{2.14}$$

This being true for every $j \in J$, we have shown that $(\overline{x}, \overline{y}, \overline{p})$ satisfies Conditions 1, 2, 3, 4 of Proposition 2.2.2.

As in the proof of Proposition 2.2.1, for each i, since $\overline{p} \cdot \overline{x}_i = \overline{p} \cdot \omega_i + \sum_{j \in J} \theta_{i,j} \overline{p} \cdot \overline{y}_j$, and letting

$$\delta_i(\overline{p}) := \{x_i \in X_i : \overline{p} \cdot x_i < \overline{p} \cdot \overline{x}_i = \overline{p} \cdot \omega_i + \sum_{j \in J} \theta_{i,j} \overline{p} \cdot \overline{y}_j\},$$

one has

$$\widehat{P}_i(\overline{x}_i) \cap \delta_i(\overline{p}) = \emptyset \iff P_i(\overline{x}_i) \cap \delta_i(\overline{p}) = \emptyset,$$

which ends the proof of Proposition 2.2.2. □

It is worth noticing with Debreu [58] that assumptions on the total production set can replace the assumptions on the individual production sets.

PROPOSITION 2.2.3 *The conclusion of Proposition 2.2.2 still holds if Assumption 2 is replaced by:*

2′ Y is convex and closed.

In counterpart, Assumption 2 implies Assumption 2′ provided that $\mathrm{A}(Y) \cap -\mathrm{A}(Y) = \{0\}$, *where for any convex set C, $\mathrm{A}(C)$ denotes the recession cone*[8] *(also called asymptotic cone) of C.*

Proof. For each j, let us replace Y_j by $Y_j' = \overline{\mathrm{co}}\, Y_j$ (the closed convex hull of Y_j) and consider the economy

$$\mathcal{E}' = \Big(\langle \mathbb{R}^\ell, \mathbb{R}^\ell\rangle, (X_i, \precsim_i, \omega_i)_{i\in I}, (Y_j')_{j\in J}, (\theta_{i,j})_{\substack{i\in I\\ j\in J}}, Z\Big).$$

The economies $\mathcal{E}$ and $\mathcal{E}'$ have same total production set. Indeed,

$$Y \subset Y' = \sum_{j\in J} \overline{\mathrm{co}}\, Y_j \subset \mathrm{cl}(\sum_{j\in J} \mathrm{co}\, Y_j) = \mathrm{cl}(\mathrm{co}(\sum_{j\in J} Y_j)) = \overline{\mathrm{co}}\, Y = Y.$$

Let then $(\overline{x}, \overline{y}', \overline{p}) \in \prod_{i\in I} X_i \times \prod_{j\in J} Y_j' \times (S \cap Z^0)$ satisfy for $\mathcal{E}'$ the conditions 1, 2, 3, 4 of Proposition 2.2.2 and $\overline{y} \in \prod_{j\in J} Y_j$ such that $\sum_{j\in J} \overline{y}_j = \sum_{j\in J} \overline{y}_j'$. As for each j, $\overline{p}\cdot\overline{y}_j \leq \overline{p}\cdot\overline{y}_j'$ together with $\overline{p}\cdot\sum_{j\in J}\overline{y}_j = \overline{p}\cdot\sum_{j\in J}\overline{y}_j'$, then $\overline{p}\cdot\overline{y}_j = \overline{p}\cdot\overline{y}_j'$ for every j and $(\overline{x}, \overline{y}, \overline{p})$ satisfies for $\mathcal{E}$ the conditions 1, 2, 3, 4.

To prove the second assertion of the proposition, let us assume 2 and that $\mathrm{A}(Y) \cap -\mathrm{A}(Y) = \{0\}$. Let $y^\nu = \sum_{j\in J} y_j^\nu$ be a sequence of elements of Y (that is, $y_j^\nu \in Y_j$) converging to y. We will distinguish two cases and prove that the second one leads to a contradiction. Indeed,

- either each sequence $(y_j^\nu)_{\nu\in\mathbb{N}}$ is bounded and, in view of Assumption 2, one can assume without loss of generality that $y_j^\nu \to y_j \in Y_j$, which proves that $y \in Y$;

[8] For the definition and properties of the recession cone of a convex set, see [157, Part II, § 8] and [87, Chapter 1, Exercises 1.17–1.21].

- or, letting $z^\nu = \max_{j\in J}\{\|y^\nu_j\|\}$, one can assume that $z^\nu \to +\infty$. Observe then that for some j_0 and an infinity of ν, $z^\nu = \|y^\nu_{j_0}\|$. Then, passing to subsequences if necessary, we have for every j, $\frac{y^\nu_j}{z^\nu} \to y_j \in \mathrm{A}(Y_j)$ and $\sum_{j\in J} y_j = 0$. It follows that $0 \neq y_{j_0} \in \mathrm{A}(Y) \cap -\mathrm{A}(Y)$, contrary to our hypothesis on Y. □

The next proposition gives conditions on the primitive characteristics of the economy insuring that Assumption 3 (b) and the second part of Assumption 1 (d) are satisfied.

PROPOSITION 2.2.4 *Assume that the individual consumption sets, the total production set and the disposal cone are convex and closed. Then*

1 $Y + Z$ *is closed provided that* $\mathrm{A}(Y) \cap (-Z) = \{0\}$*;*

2 the attainable consumption sets are compact provided the two following conditions hold:

(a) $\mathrm{A}(\sum_{i\in I} X_i) \cap \mathrm{A}(Y + Z) = \{0\}$,

(b) $\mathrm{A}(\sum_{i\in I} X_i) \cap (-\mathrm{A}(\sum_{i\in I} X_i)) = \{0\}$.

3 If, in addition, $\mathrm{A}(Y) \cap (-\mathrm{A}(Y)) = \{0\}$*, then the attainable production sets* $\widehat{Y}_j$ *are relatively compact.*

Proof. To prove 1, assume $\mathrm{A}(Y)\cap(-Z) = \{0\}$ and that $(y^\nu + z^\nu)_{\nu\in\mathbb{N}}$ is a sequence of elements of $Y + Z$ converging to t. If the sequences $(y^\nu)_{\nu\in\mathbb{N}}$ and $(z^\nu)_{\nu\in\mathbb{N}}$ are bounded, one can assume that $y^\nu \to y \in Y$, $z^\nu \to z \in Z$, which proves that $t = y + z \in Y + Z$. If not, one can assume that the sequence $v^\nu = \max\{\|y^\nu\|, \|z^\nu\|\}$ tends to $+\infty$. Then either $v^\nu = \|y^\nu\|$ for an infinity of ν or $v^\nu = \|z^\nu\|$ for an infinity of ν. In both cases, one can assume that $\frac{y^\nu}{v^\nu} \to y \in \mathrm{A}(Y)$, $\frac{z^\nu}{v^\nu} \to z \in \mathrm{A}(Z) = Z$, with $y + z = 0$, $y \neq 0$, $z \neq 0$, a contradiction with the hypothesis.

The proof of 2 is similar. Let for $i \in I$, $(x^\nu_i)_{\nu\in\mathbb{N}}$ be a sequence of elements of $\widehat{X}_i$. To each x^ν_i we associate $x^\nu_{i'} \in X_i$, $i' \neq i$ and $t^\nu \in Y + Z$ such that

$$x^\nu_i + \sum_{i'\neq i} x^\nu_{i'} = \omega + t^\nu$$

and prove by contraposition that x^ν_i and all $x^\nu_{i'}$ are bounded. If not, let $v^\nu = \max\{\|x^\nu_i\|, (\|x^\nu_{i'}\|)_{i'\neq i}\}$. Without loss of generality one can assume that $v^\nu \to +\infty$, and that $\frac{x^\nu_i}{v^\nu} \to x_i \in \mathrm{A}(X_i)$, each $\frac{x^\nu_{i'}}{v^\nu} \to x_{i'} \in \mathrm{A}(X_{i'})$, $\frac{t^\nu}{v^\nu} \to t \in \mathrm{A}(Y + Z)$, with $t = x_i + \sum_{i'\neq i} x_{i'}$. Observe that

$t \in \mathrm{A}(Y+Z) \cap \mathrm{A}(\sum_{i \in I} X_i)$, thus that $x_i + \sum_{i' \neq i} x_{i'} = 0$. On the other hand, either x_i or one of the $x_{i'}$ is different from 0, which contradicts the hypothesis $\mathrm{A}(\sum_{i \in I} X_i) \cap (-\mathrm{A}(\sum_{i \in I} X_i)) = \{0\}$.

We now deduce from the fact that a subsequence of $(x_i^\nu)_{\nu \in \mathbb{N}}$ converges to $x_i \in X_i$ that $\widehat{X}_i$ is compact. The same proof can be done for any $i \in I$.

For proving 3, for $j \in J$, let us consider a sequence $(y_j^\nu)_{\nu \in \mathbb{N}}$ of elements of $\widehat{Y}_j$. To each y_j^ν we associate $y_{j'}^\nu \in Y_j$, $j' \neq j$, $x_i^\nu \in X_i$, $i \in I$, and $z^\nu \in Z$ such that

$$\sum_{i \in I} x_i^\nu = \omega + y_j^\nu + \sum_{j' \neq j} y_{j'}^\nu + z^\nu$$

and prove by contraposition that y_j^ν and all $y_{j'}^\nu$ are bounded. If not, let $v^\nu = \max\{\|y_j^\nu\|, (\|y_{j'}^\nu\|)_{j' \neq j}\}$. Without loss of generality one can assume that $v^\nu \to +\infty$, and, in view of 2, that $\frac{y_j^\nu}{v^\nu} \to y_j \in \mathrm{A}(Y_j)$, each $\frac{y_{j'}^\nu}{v^\nu} \to y_{j'} \in \mathrm{A}(Y_{j'})$, with $0 = y_j + \sum_{j' \neq j} y_{j'}$. As either y_j or one of the $y_{j'}$ is different from 0, and since for each j, $\mathrm{A}(Y_j) \subset \mathrm{A}(Y)$, one gets $\mathrm{A}(Y) \cap (-\mathrm{A}(Y)) \neq \{0\}$, in contradiction with the hypothesis. □

The (eventual) assumption $\mathrm{A}(Y) \cap -\mathrm{A}(Y) = \{0\}$ is a variant of the irreversibility of total production. As emphasized by McKenzie [136], [137], Debreu [59], this assumption is superfluous at least for existence of transitive equilibrium. In the next chapter, we will see that it is no longer true when one deals with dependent consumer's preferences. In counterpart, to assume that the attainable production sets are compact will simplify the limiting process which allows for passing from quasiequilibrium existence in a compact economy to quasiequilibrium existence in the original economy.

Recalling that $Z \subset -\mathbb{R}_+^\ell$, the assumption $\mathrm{A}(Y) \cap (-Z) = \{0\}$ is a weak variant of the impossibility of free production. Actually the lower boundedness of the consumption sets and the impossibility of free production are respectively replaced by $\mathrm{A}(\sum_{i \in I} X_i) \cap (-\mathrm{A}(\sum_{i \in I} X_i)) = \{0\}$ and $\mathrm{A}(\sum_{i \in I} X_i) \cap \mathrm{A}(Y+Z) = \{0\}$, yielding a coordinate-free version of these classical hypothesis.

Notice finally that every quasiequilibrium of $\mathcal{E}$, in particular the quasiequilibrium proved to exist in Proposition 2.2.2, is nontrivial[9] under a simple sufficient condition.

---

[9] As defined in Definition 2.0.5, a quasiequilibrium $(\overline{x}, \overline{y}, \overline{p})$ is nontrivial if and only if $\delta_i(\overline{p}) \neq \emptyset$ for some $i \in I$.

PROPOSITION 2.2.5 *Every quasiequilibrium is nontrivial provided that*

- $\omega \in \text{int}(\sum_{i\in I} X_i - Y - Z)$.

Proof. Let $(\overline{x}, \overline{y}, \overline{p})$ be a quasiequilibrium of $\mathcal{E}$. Recalling that $\overline{p} \neq 0$, pick $u \in \mathbb{R}^\ell$ such that $\overline{p} \cdot u < 0$ and, in view of the above condition, $(x, y) \in \prod_{i\in I} X_i \times \prod_{j\in J} Y_j$ such that $\omega + u \in \sum_{i\in I} X_i - Y - Z$. This shows that

$$\overline{p} \cdot \sum_{i\in I}(x_i - \omega_i) < \overline{p} \cdot \sum_{j\in J} y_j \leq \overline{p} \cdot \sum_{j\in J} \overline{y}_j = \sum_{i\in I}\sum_{j\in J} \theta_{i,j}\overline{p} \cdot \overline{y}_j,$$

thus that for some $i \in I$, $\delta_i(\overline{p}) \neq \emptyset$. □

2.3 From quasiequilibrium to equilibrium

In order to be able to use the results of this section in the next chapters, let us now come back to the general framework presented in the introduction to this chapter, where consumers' preferences are described by preference correspondences $P_i : \prod_{k\in I} X_k \times \prod_{j\in J} Y_j \times S \to X_i$ satisfying $x_i \notin P_i(x, y, p)$. We thus start with a quasiequilibrium $(\overline{x}, \overline{y}, \overline{p})$, as defined in Definition 2.0.5, of the economy

$$\mathcal{E} = \Big(\langle \mathbb{R}^\ell, \mathbb{R}^\ell\rangle, (X_i, P_i, \omega_i)_{i\in I}, (Y_j)_{j\in J}, (\theta_{i,j})_{\substack{i\in I\\ j\in J}}, Z\Big).$$

Showing that a quasiequilibrium of $\mathcal{E}$ is actually an equilibrium requires assumptions which are suggested by the following proposition.

PROPOSITION 2.3.1 *Assume that each X_i is convex and*

1. *For all $i \in I$, for all $(x, y, p) \in \mathcal{A}(E) \times (S \cap Z^0)$, if $z_i \in P_i(x, y, p)$ and $v_i \in X_i$, there exists λ: $0 < \lambda \leq 1$ such that $(\lambda v_i + (1-\lambda)z_i) \in P_i(x, y, p)$.*

Then $x_i \in P_i(\overline{x}, \overline{y}, \overline{p}) \Longrightarrow \overline{p} \cdot x_i > \overline{p} \cdot \overline{x}_i$, for each $i \in I$ such that $\delta_i(\overline{p}) \neq \emptyset$.

Proof. Let $i \in I$ be such that $\delta_i(\overline{p}) \neq \emptyset$. Pick $v_i \in \delta_i(\overline{p})$ and $x_i \in P_i(\overline{x}, \overline{y}, \overline{p})$. By definition of quasiequilibrium, $\overline{p} \cdot x_i \geq \overline{p} \cdot \overline{x}_i$ and $(\overline{x}, \overline{y}, \overline{p}) \in \mathcal{A}(E) \times (S \cap Z^0)$. Applying the above assumption, let λ: $0 < \lambda \leq 1$ be such that $(\lambda v_i + (1-\lambda)x_i) \in P_i(\overline{x}, \overline{y}, \overline{p})$. If $\overline{p} \cdot x_i = \overline{p} \cdot \overline{x}_i$, then $(\lambda v_i + (1-\lambda)x_i) \in P_i(\overline{x}, \overline{y}, \overline{p}) \cap \delta_i(\overline{p})$, in contradiction with $P_i(\overline{x}, \overline{y}, \overline{p}) \cap \delta_i(\overline{p}) = \emptyset$. Consequently, $\overline{p} \cdot x_i > \overline{p} \cdot \overline{x}_i$. □

REMARK 2.3.1 Condition 1 of Proposition 2.3.1, borrowed from Bergstrom [30], is satisfied if for every $(x, y, p) \in \mathcal{A}(\mathcal{E}) \times (S \cap Z^0)$, and for each i, X_i is convex and the set $P_i(x, y, p)$ is open in X_i. Condition 1 of Proposition 2.3.1 is also satisfied if for each i, X_i is convex and P_i is defined by a concave utility function on X_i. Moreover, as it is easily verified, Condition 1 of Proposition 2.3.1 is inherited by the "augmented preference correspondences"

$$\widehat{P}_i(x, y, p) = \{x'_i \in X_i : x'_i = x_i + \lambda(x''_i - x_i), 0 < \lambda \leq 1,\ x''_i \in P_i(x, y, p)\}$$

provided it is satisfied by correspondences P_i.

Under either condition, at a quasiequilibrium $(\overline{x}, \overline{y}, \overline{p})$, each consumer i for whom $\delta_i(\overline{p}) \neq \emptyset$ is actually at his own equilibrium. It then suffices to add conditions insuring that $\delta_i(\overline{p}) \neq \emptyset$ for each $i \in I$ for proving that $(\overline{x}, \overline{y}, \overline{p})$ is actually an equilibrium. A simple way of obtaining such a result is given in the following proposition.

PROPOSITION 2.3.2 *Let $(\overline{x}, \overline{y}, \overline{p})$ be a quasiequilibrium of $\mathcal{E}$. Assume that each X_i is convex and*

1 *For all $i \in I$, for all $(x, y, p) \in \mathcal{A}(E) \times (S \cap Z^0)$, if $z_i \in P_i(x, y, p)$ and $v_i \in X_i$, there exists λ: $0 < \lambda \leq 1$ such that $(\lambda v_i + (1 - \lambda) z_i) \in P_i(x, y, p)$;*

2 *For all $i \in I$, $\omega_i \in \operatorname{int}(X_i - \sum_{j \in J} \theta_{i,j} Y_j - Z)$.*

Then $\delta_i(\overline{p}) \neq \emptyset$ for each $i \in I$ and $(\overline{x}, \overline{y}, \overline{p})$ is an equilibrium of $\mathcal{E}$.

Proof. Recalling that $\overline{p} \neq 0$, for each $i \in I$, pick $u \in \mathbb{R}^\ell$ such that $\overline{p} \cdot u < 0$ and, in view of Condition 2, $(\omega_i + u) \in (X_i - \sum_{j \in J} \theta_{i,j} Y_j - Z)$. Thus for some $x_i \in X_i$, $\overline{p} \cdot x_i < \overline{p} \cdot \omega_i + \sum_{j \in J} \theta_{i,j} \overline{p} \cdot \overline{y}_j$ and $\delta_i(\overline{p}) \neq \emptyset$. □

In what follows, we weaken Condition 2, starting now from a nontrivial quasiequilibrium, and we set on the other side an assumption of irreducibility of the economy, first formulated by Gale [92] in the context of a linear exchange economy. The general idea of this structural hypothesis is to assume, at any couple of an attainable allocation and of a price in $S \cap Z^0$, a possible dependence of the welfare of any group of agents on the resources of the agents who do not belong to this group. Such an interdependence is easily verified in simple contexts. In the general framework of this section, we set the following definitions adapted from McKenzie [136], [137], [138], Debreu [59], Arrow–Hahn [22], Bergstrom [30], Hammond [101], and Florig [89].

From now, we assume that the total production set Y and the individual consumption sets are convex and that $0 \in Y$. If $y \in \prod_{j \in J} Y_j$, define $\underline{y} = \sum_{j \in J} y_j$ and let $T_{\underline{y}}(Y)$ and $N_{\underline{y}}(Y)$ denote respectively the tangent cone and the normal cone of Y at the point $\underline{y}$:

$$T_{\underline{y}}(Y) := \mathrm{cl}\{z \in \mathbb{R}^{\ell} : z = \lambda(y' - \underline{y}),\ \lambda > 0,\ y' \in Y\},$$

$$N_{\underline{y}}(Y) := (T_{\underline{y}}(Y))^0.$$

DEFINITION 2.3.1 *$\mathcal{E}$ is* ***McKenzie–Debreu irreducible*** *if for any partition $\{I_1, I_2\}$ of the set of consumers into two nonempty subsets and for each $(x, y, p) \in \mathcal{A}(\mathcal{E}) \times (S \cap Z^0)$, there exists $x' \in \prod_{i \in I} X_i$ satisfying the following conditions:*

1 $x'_i \in \mathrm{cl}(\widehat{P}_i(x, y, p))$ for each $i \in I_1$, with for some $i_1 \in I_1$, $x'_{i_1} \in \widehat{P}_{i_1}(x, y, p)$,

2 $\sum_{i \in I_1}(x'_i - x_i) + \sum_{i \in I_2}(x'_i - \omega_i) \in (T_{\underline{y}}(Y) + Z)$.

In other words, $\mathcal{E}$ is **McKenzie–Debreu irreducible** if for any partition $\{I_1, I_2\}$ of the set of consumers into two nonempty subsets and for each couple of an attainable allocation $(x, y) \in \mathcal{A}(E)$ and of a price vector $p \in S \cap Z^0$, the group I_1 (after a possible disposal) may be moved to a preferred position, as expressed by 1, by adding a vector of the tangent cone of Y at $\underline{y}$ plus a feasible trade from I_2. In the case of strictly convex complete preference preorders $\succsim_i$ on X_i satisfying the continuity assumption 1 (b) of Section 2.2, for each i, $\widehat{P}_i(x, y, p) = P_i(x_i) = \{x'_i \in X_i : x'_i \succ_i x_i\}$ and $\mathrm{cl}\,\widehat{P}_i(x, y, p) = \{x'_i \in X_i : x'_i \succsim_i x_i\}$. Then Definition 2.3.1 coincides with McKenzie [138]'s formulation of irreducibility.

DEFINITION 2.3.2 *$\mathcal{E}$ is* ***Arrow–Hahn irreducible*** *if for any partition $\{I_1, I_2\}$ of the set of consumers into two nonempty subsets and for each $(x, y, p) \in \mathcal{A}(\mathcal{E}) \times (S \cap Z^0)$, there exists $x' \in \prod_{i \in I} X_i$ and $\omega' \in \mathbb{R}^{\ell}$ satisfying the following conditions:*

1 $x'_i \in \mathrm{cl}(\widehat{P}_i(x, y, p))$ for each $i \in I_1$, with for some $i_1 \in I_1$, $x'_{i_1} \in \widehat{P}_{i_1}(x, y, p)$,

2 $(\sum_{i\in I} x'_i - \omega') \in (\{\underline{y}\} + T_{\underline{y}}(Y) + Z)$ *with, for each* $h = 1, 2, \ldots, \ell$,

(a) $\omega'^h > \omega^h \Longrightarrow \exists\lambda_h > 0, \sum_{i\in I_2} \omega_i - \lambda_h e^h \in \sum_{i\in I_2} X_i - Z$

(b) $\omega'^h < \omega^h \Longrightarrow \exists\lambda_h > 0, \sum_{i\in I_2} \omega_i + \lambda_h e^h \in \sum_{i\in I_2} X_i - Z.$

The interpretation of this definition is that, for any partition $\{I_1, I_2\}$ of the set of consumers into two nonempty subsets and for each couple of an attainable allocation $(x, y) \in \mathcal{A}(E)$ and of a price vector $p \in S \cap Z^0$, the group I_1 (after a possible disposal) may be moved to a preferred position by adding to $\underline{y}$ a vector of the tangent cone of Y at $\underline{y}$ and increasing (resp. decreasing) the total resource of commodities which can be supplied in positive (resp. negative) amount by the group I_2.

If free-disposal prevails, and for continuous and strictly convex complete preference preorders, Arrow-Hahn [22] actually give definitions that we next rewrite in our framework.

DEFINITION 2.3.3 *In the economy* $\mathcal{E}$, *consumer* i_1 *is said to be* ***resource related*** *to consumer* i_2 *if for each* $(x, y, p) \in \mathcal{A}(\mathcal{E}) \times (S \cap Z^0)$, *there exists* $x' \in \prod_{i\in I} X_i$ *and* $\omega' \in \mathbb{R}^\ell$ *satisfying the following conditions:*

1 $x'_i \in \mathrm{cl}(\widehat{P}_i(x, y, p))$ *for each* $i \in I$, *with* $x'_{i_1} \in \widehat{P}_{i_1}(x, y, p)$,

2 $(\sum_{i\in I} x'_i - \omega') \in (\{\underline{y}\} + T_{\underline{y}}(Y) + Z)$ *with, for each* $h = 1, 2, \ldots, \ell$,

(a) $\omega'^h > \omega^h \Longrightarrow \exists\lambda_h > 0, \omega_{i_2} - \lambda_h e^h \in X_{i_2} - Z,$

(b) $\omega'^h < \omega^h \Longrightarrow \exists\lambda_h > 0, \omega_{i_2} + \lambda_h e^h \in X_{i_2} - Z.$

DEFINITION 2.3.4 *In the economy* $\mathcal{E}$, *consumer* i *is said to be* ***indirectly resource related*** *to consumer* i' *if there is a sequence of consumers* $i_0, \ldots, i_p$ *such that* $i_0 = i$, $i_p = i'$ *and* i_k *is resource related to* i_{k+1} $(k = 0, \ldots, p-1)$.

It is readily seen that an economy where every consumer is indirectly resource related to every other is Arrow–Hahn irreducible.

The next definition is more abstract. For an exchange economy, it reduces to what is called Bergstrom irreducibility in [78].

DEFINITION 2.3.5 *$\mathcal{E}$ is **Bergstrom–Florig irreducible** if for any partition $\{I_1, I_2\}$ of the set of consumers into two nonempty subsets and for each $(x, y, p) \in \mathcal{A}(\mathcal{E}) \times (S \cap Z^0)$, there exist real numbers $\theta_i > 0$, $i \in I$, and $x' \in \prod_{i \in I} X_i$ satisfying the following conditions:*

1 $x'_i \in \mathrm{cl}(\widehat{P}_i(x, y, p))$ for each $i \in I_1$, with for some $i_1 \in I_1$, $x'_{i_1} \in \widehat{P}_{i_1}(x, y, p)$,

2 $\sum_{i \in I} \theta_i(x'_i - \omega_i - \sum_{j \in J} \theta_{i,j} y_j) \in (T_{\underline{y}}(Y) + Z)$.

We will see in the corollaries 2.3.2 and 2.3.3 to the next proposition that Mckenzie–Debreu irreducibility and Arrow–Hahn irreducibility independently imply Bergstrom–Florig irreducibility. The important observation here is that the Bergstrom–Florig irreducibility is a sufficient condition for passing from nontrivial quasiequilibrium to equilibrium. When preferences do not depend on prices, this condition is also necessary provided that every quasi-equilibrium is nontrivial.

Notice incidentally that any irreducibility assumption implies the insatiability of each consumer at every $(x, y, p) \in \mathcal{A}(\mathcal{E}) \times (S \cap Z^0)$.

PROPOSITION 2.3.3 *Assume that each sets X_i and Y_j are convex. Let $(\overline{x}, \overline{y}, \overline{p})$ be a nontrivial quasiequilibrium of $\mathcal{E}$. Then $(\overline{x}, \overline{y}, \overline{p})$ is an equilibrium if:*

1 For all $i \in I$, for all $(x, y, p) \in \mathcal{A}(\mathcal{E}) \times (S \cap Z^0)$, if $z_i \in P_i(x, y, p)$ and $v_i \in X_i$, there exists $\lambda: 0 < \lambda \leq 1$ such that $(\lambda v_i + (1 - \lambda) z_i) \in P_i(x, y, p)$;

2 $\mathcal{E}$ is Bergstrom–Florig irreducible.

Proof. Let $I_1 = \{i \in I : \delta_i(\overline{p}) \neq \emptyset\}$ and $I_2 = \{i \in I : \delta_i(\overline{p}) = \emptyset\}$. Since the quasiequilibrium $(\overline{x}, \overline{y}, \overline{p})$ is assumed to be nontrivial, $I_1 \neq \emptyset$.

Assume now that $I_2 \neq \emptyset$. Recalling that $(\overline{x}, \overline{y}, \overline{p}) \in \mathcal{A}(\mathcal{E}) \times (S \cap Z^0)$, let $\theta_i > 0$, $i \in I$ and $x' \in \prod_{i \in I} X_i$ satisfy for the partition $\{I_1, I_2\}$ the conditions of Definition 2.3.5. Using the quasiequilibrium definition for the agents of I_1 and the definition of I_2, one has for every $i \in I$

$$\theta_i(\overline{p} \cdot (x'_i - \omega_i)) \geq \theta_i \overline{p} \cdot \sum_{j \in J} \theta_{i,j} \overline{y}_j$$

with a strict inequality for some $i \in I_1$. Summing over i, one gets

$$\overline{p} \cdot \sum_{i \in I} \theta_i (x'_i - \omega_i - \sum_{j \in J} \theta_{i,j} \overline{y}_j) > 0.$$

On the other hand, it follows from Condition 2 in Definition 2.3.5 and again the definition of quasiequilibrium that

$$\overline{p} \cdot \sum_{i \in I} \theta_i (x'_i - \omega_i - \sum_{j \in J} \theta_{i,j} \overline{y}_j) \leq 0,$$

which yields a contradiction.

We finally conclude using Proposition 2.3.1. □

COROLLARY 2.3.1 (BERGSTROM) *Let $\mathcal{E}$ be an exchange economy (with disposal activities) and individual preferences which do not depend on prices. The conclusion of Proposition 2.3.3 still holds if Condition 2 is replaced by:*

If $x \in \prod_{i \in I} X_i$ is an attainable allocation, then for each i there exists $x' \in \prod_{i \in I} X_i$ such that $x'_{i'} \in P_{i'}(x)$ $\forall i' \neq i$ and $\sum_{i' \neq i} (x'_{i'} - \omega_{i'}) + \theta_i (x_i - \omega_i) \in Z$ for some scalar $\theta_i > 0$.

Proof. The proof is obvious. □

COROLLARY 2.3.2 (MCKENZIE–DEBREU) *The conclusion of Proposition 2.3.3 still holds if Condition 2 is replaced by*

- *$\mathcal{E}$ is McKenzie–Debreu irreducible.*

Proof. It suffices to prove that a McKenzie–Debreu irreducible economy is Bergstrom–Florig irreducible. Let $\{I_1, I_2\}$ be a partition of the set of agents into nonempty subsets of I and $(x, y, p) \in \mathcal{A}(\mathcal{E}) \times (S \cap Z^0)$. Assume that $x \in \prod_{i \in I} X_i$ satisfies for the partition $\{I_1, I_2\}$ the conditions of Definition 2.3.1. From the relation

$$\sum_{i \in I_1} (x'_i - x_i) + \sum_{i \in I_2} (x'_i - x_i) + \sum_{i \in I_2} (x_i - \omega_i) \in T_{\underline{y}}(Y) + Z,$$

one gets

$$\sum_{i \in I} (x'_i - \omega_i - \sum_{j \in J} \theta_{i,j} y_j) + \sum_{i \in I_2} (x_i - \omega_i) \in (T_{\underline{y}}(Y) + Z),$$

which implies

$$\sum_{i\in I_1}(x'_i - \omega_i - \sum_{j\in J}\theta_{i,j}y_j) + 2\sum_{i\in I_2}(\frac{x'_i + x_i}{2} - \omega_i - \sum_{j\in J}\theta_{i,j}y_j)$$

$$\in (\{-\sum_{i\in I_2}\sum_{j\in J}\theta_{i,j}y_j\} + T_{\underline{y}}(Y) + Z).$$

It is readily verified that $-\sum_{i\in I_2}\sum_{j\in J}\theta_{i,j}y_j \in T_{\underline{y}}(Y)$. In view of the convexity of each X_i , this proves Condition 2 of Definition 2.3.5. □

COROLLARY 2.3.3 (ARROW–HAHN) *The conclusion of Proposition 2.3.3 still holds if Condition 2 is replaced by*

- *$\mathcal{E}$ is Arrow–Hahn irreducible.*

Proof. Let us prove that an Arrow–Hahn irreducible economy is Bergstrom–Florig irreducible. Let $\{I_1, I_2\}$ be a partition of the set of agents into nonempty subsets of I and $(x, y, p) \in \mathcal{A}(\mathcal{E}) \times (S \cap Z^0)$. Assume that $x \in \prod_{i\in I} X_i$ satisfies for the partition $\{I_1, I_2\}$ the conditions of Definition 2.3.2. Setting

$$\omega' - \omega = \sum_{h=1}^{\ell}\mu_h e^h = \sum_{\mu<0}\frac{|\mu_h|}{\lambda_h}(-\lambda_h e^h) + \sum_{\mu>0}\frac{\mu_h}{\lambda_h}(\lambda_h e^h),$$

letting

$$\alpha = \sum_{\mu<0}\frac{|\mu_h|}{\lambda_h} + \sum_{\mu>0}\frac{\mu_h}{\lambda_h}$$

and using the convexity of $\sum_{i\in I_2}(\{\omega_i\} - X_i)$, we get

$$\omega' - \omega \in (\alpha\sum_{i\in I_2}(\{\omega_i\} - X_i) + Z).$$

Let for each $i \in I_2$, x''_i be such that $\omega' - \omega + \alpha\sum_{i\in I_2}(x''_i - \omega_i) \in Z$. We then have

$$\sum_{i\in I_1}(x'_i - \omega_i - \sum_{j\in J}\theta_{i,j}y_j) + (1+\alpha)\sum_{i\in I_2}(\frac{x'_i + \alpha x''_i}{1+\alpha} - \omega_i - \sum_{j\in J}\theta_{i,j}y_j)$$

$$\in (\{-\alpha\sum_{i\in I_2}\sum_{j\in J}\theta_{i,j}y_j\} + T_{\underline{y}}(Y) + Z).$$

Once again, it is readily verified that $-\alpha \sum_{i \in I_2} \sum_{j \in J} \theta_{i,j} y_j \in T_{\overline{y}}(Y)$. Using the convexity of each X_i , this proves Condition 2 of Definition 2.3.5.

□

The necessity of Bergstrom–Florig irreducibility when preferences do not depend on prices is precisely expressed in the next proposition.

PROPOSITION 2.3.4 (FLORIG) *Assume that Y and each X_i are convex and that each P_i: $\prod_{k \in I} X_k \times \prod_{j \in J} Y_j \to X_i$ is convex valued with nonempty values for each $(x, y) \in \mathcal{A}(\mathcal{E})$. If $\mathcal{E}$ is not Bergstrom–Florig irreducible, then $\mathcal{E}$ has a quasiequilibrium (x, y, p) such that for some $i \in I$, $\delta_i(p) = \emptyset$.*

Proof. Let (x, y) and $\{I_1, I_2\}$ be a partition of the set of consumers into two nonempty subsets of I such that Conditions 1 and 2 of Definition 2.3.5 are not simultaneously verified. One then has: $A \cap (T_{\underline{y}}(Y) + Z) = \emptyset$, where

$$A = \left\{ \sum_{i \in I} \theta_i (x'_i - \omega_i - \sum_{j \in J} \theta_{i,j} y_j) : \begin{array}{c} x'_i \in \widehat{P}_i(x, y) \ \forall i \in I_1, \\ x'_i \in X_i \ \forall i \in I_2, \\ \theta_i > 0 \ \forall i \in I \end{array} \right\}.$$

In view of the convexity of the sets $(\widehat{P}_i(x, y) - \omega_i - \sum_{j \in J} \theta_{i,j} y_j)$, $i \in I_1$ and $(X_i - \omega_i - \sum_{j \in J} \theta_{i,j} y_j)$, $i \in I_2$, A is the convex cone generated by the union of these sets. Let $p \in S$ which separates the two convex cones A and $(T_{\underline{y}}(Y) + Z)$. Since $p \cdot (\widehat{P}_i(x, y) - \omega_i - \sum_{j \in J} \theta_{i,j} y_j) \geq 0$ for each $i \in I_1$, $p \cdot (X_i - \omega_i - \sum_{j \in J} \theta_{i,j} y_j) \geq 0$ for each $i \in I_2$, while $p \cdot (T_{\underline{y}}(Y) + Z) \leq 0$, (x, y, p) is a quasiequilibrium of $\mathcal{E}$ such that $\delta_i(p) = \emptyset$ for each $i \in I_2$.

□

Chapter 3

NONTRANSITIVE EQUILIBRIUM OF A FINITE DIMENSIONAL COMPETITIVE ECONOMY

In this book, nontransitive equilibrium is negatively defined by the fact that consumers are not assumed to have a complete preference preorder on their consumption set.

This purely negative definition leads to include first in non transitive equilibrium the case where the preference preorders of the consumers on their consumption set are not complete, that is, the case where the potential choices of consumers have the strong coherence associated with transitivity of partial preorders but where consumers cannot preorder all the couples of commodity bundles proposed to their consumption. Equivalently, if one is only given transitive strict preference relations, the indifference of consumers between two possible consumptions need not coincide with the equivalence relations (by definition transitive) naturally associated with their transitive strict preference relations (for this construction, see Fishburn [76]).

Nontransitive equilibrium corresponds also to the case where the strict binary preference relations of consumers on their consumption set are the only data on consumers' preferences and where these relations are not transitive. Equivalently, the preferences of each consumer i on his consumption set X_i are described by a correspondence P_i: $X_i \to X_i$ satisfying $x_i \notin P_i(x_i)$.

Finally, to replace complete preference preorders by preference correspondences leads to include in nontransitive equilibrium the case where consumers have preferences over their consumption set with regard to the actions of the other agents and to relative prices, as in the general equilibrium model described in the preamble to Chapter 2. One should notice however that considering externalities in preferences is not related with the absence of transitivity. Game theorists use for each agent utility functions associating a real number to a t-uple of simultaneous strategies of all players. In general equilibrium, McKenzie [135] and Arrow-Hahn [22] have considered transitive dependent preferences.[1]

This chapter addresses the equilibrium existence problem in the general framework of an economy

$$\mathcal{E} = \Big(\langle \mathbb{R}^\ell, \mathbb{R}^\ell \rangle, (X_i, P_i, \omega_i)_{i \in I}, (Y_j)_{j \in J}, (\theta_{i,j})_{\substack{i \in I \\ j \in J}}, Z\Big)$$

where the preferences of each consumer i on his consumption set X_i are described by a correspondence P_i: $\prod_{i \in I} X_i \times \prod_{j \in J} Y_j \times S \to X_i$ satisfying $x_i \notin P_i(x, y, p)$.

As already stressed in the preamble to Chapter 2, the excess demand approach used in Chapter 2 is no longer applicable to the previous contexts without a strengthening of assumptions that we want to avoid. With the exception of the seminal paper of Mas-Colell [126] and of a first paper of Shafer–Sonnenschein [167], the equilibrium existence in the model we deal with is implicitly or explicitly based on a previous equilibrium existence result for an "abstract economy", a term introduced by Debreu [55] for defining a generalized game. Applying this result to the economic model consists in considering the equilibrium functioning of an economy as the equilibrium of a generalized game where each traditional agent (consumer or producer) taking prices as given uses his conventional optimizing under constraints behavior while, according to a commonly held view of the role of prices, an additional agent, a kind of price setting agency, corrects the eventual disequilibrium by making the excess of demand over supply as expensive as possible.

This simultaneous optimization approach was the one used in Arrow–Debreu [21], the paper which introduced the Arrow–Debreu model and first proved its consistency. However, in order to prove the equilibrium of an abstract economy, we shall not use the generalization by Debreu [55] of the Nash theorem since the equilibrium existence proof in this paper is based on the convexity of values of correspondences which are nothing else than the demand correspondence when the Debreu result is applied

[1] A warning is also necessary. Not all the effects of the economic behavior of others are mediated through their influence on preferences of the agents. One should also consider that consumption and production sets depend on the current vector of actions of other agents. Such a complete study of externalities may lead to nonconvexities whose consideration is out of the scope of this chapter.

to the abstract economy associated with an economy. The existence of equilibrium without this assumption of convexity of demand correspondences is the fundamental contribution of Gale and Mas-Colell ([94] and [96]).

The chapter is organized as follows. In Section 3.1 we introduce the concept of abstract economy and define its equilibrium. In order to be able to drop the assumptions of strong autarky of consumers in the private ownership economy to which we will associate the abstract economy, we also define a concept of quasiequilibrium for an abstract economy in such a way that a quasiequilibrium of an abstract economy be associated with a quasiequilibrium of the private ownership economy. Then we give sufficient conditions for its existence.

In Section 3.2, we prove the existence of nontransitive quasiequilibrium. As in Chapter 2, we first compactify the economy and apply to the compactified economy the results of the previous section. Doing this, in order to get the best possible results, we use the same tricks as in Chapter 2. We then prove that a quasiequilibrium of the compact economy is also a quasiequilibrium of the original economy.

Sufficient conditions for passing from the existence of a quasiequilibrium to the existence of an equilibrium were already given in Section 2.3 of Chapter 2 for our general model. We refer the reader to this section.

3.1 Equilibrium and quasiequilibrium of an abstract economy

An *abstract economy* (or *generalized qualitative game*) is completely specified by

$$\Gamma = ((X_i, \alpha_i, P_i)_{i \in N})$$

where N is a finite set of agents (players) and for each $i \in N$,

- X_i is a *choice set* (or *strategy set*) that we will assume in this chapter to be a subset of some finite dimensional vector space. As N is finite, we can assume that $\mathbb{R}^\ell$ is the ambient space for some $\ell \in \mathcal{N}$ and that, for every i, $X_i \subset \mathbb{R}^\ell$.
- The correspondence
$$\alpha_i : \prod_{k \in N} X_k \to X_i$$
is called *constraint correspondence* (or *feasible strategy correspondence*). Let $X := \prod_{i \in N} X_i$; for each $x \in X$, $\alpha_i(x)$ is interpreted a

set of possible strategies for i, given the choices $(x_j)_{j \neq i}$ of the other agents.

- The correspondence
$$P_i : \prod_{k \in N} X_k \to X_i$$
is a *preference correspondence.* For each $x \in X$, under the condition $x_i \notin P_i(x)$, $P_i(x)$ is interpreted as the set of elements of X_i strictly preferred by i to x_i when the choice of the other agents is $(x_j)_{j \neq i}$.

A particular case is the one where the preferences of each agent i are defined by a utility function $u_i\colon X \to \mathbb{R}$. Using the notations $x_{-i} := (x_{i'})_{i' \neq i} \in \prod_{i' \neq i} X_{i'}$ and $u_i : (X_{-i} \times X_i) \to \mathbb{R}$, then

$$P_i(x) = \{x'_i : u_i(x_{-i}, x'_i) > u_i(x_{-i}, x_i)\}.$$

The abstract economy $\Gamma = ((X_i, \alpha_i, u_i)_{i \in N})$ is for game theorists a *generalized game.* If moreover for every $i \in N$, for every $x \in X$, $\alpha_i(x) = X_i$, then Γ is a classical game in normal form.

DEFINITION 3.1.1 *Let $\alpha : X \to X$ be defined by $\alpha(x) = \prod_{k \in N} \alpha_k(x)$.*

$$\mathcal{A}(\Gamma) = \{x \in X\colon\ x \in \alpha(x)\},$$

the set of fixed points of the correspondence α, is the ***set of feasible strategy vectors.***

For abstract economies, a natural extension of the notion of Nash equilibrium of a game is the following.

DEFINITION 3.1.2 *A t-uple of strategies $\overline{x} = (\overline{x}_i)_{i \in N} \in X$ is an* ***equilibrium*** *of the abstract economy Γ if for every $i \in N$,*

1 $\overline{x}_i \in \alpha_i(\overline{x})$

2 $P_i(\overline{x}) \cap \alpha_i(\overline{x}) = \emptyset.$

The first condition states that $\overline{x}$ is a feasible strategy vector. The second condition expresses that each agent has chosen in the set defined by his constraint correspondence a feasible strategy which best satisfies his preferences.

For each $i \in N$, let now $\beta_i\colon \prod_{k \in N} X_k \to X_i$ be a correspondence satisfying for all $x \in \prod_{k \in N} X_k$

$$\beta_i(x) \subset \alpha_i(x); \tag{3.1}$$

$$\text{if } \beta_i(x) \neq \emptyset, \text{ then } \text{cl}(\beta_i(x)) = \text{cl}(\alpha_i(x)). \tag{3.2}$$

DEFINITION 3.1.3 *Given $\beta = (\beta_i)_{i \in N}$ as above, $\overline{x} = (\overline{x}_i)_{i \in N} \in X$ is a **β-quasiequilibrium** of Γ if for each $i \in N$,*

1 $\overline{x}_i \in \alpha_i(\overline{x})$

2 $P_i(\overline{x}) \cap \beta_i(\overline{x}) = \emptyset$.

We will see later that, applied to the abstract economy associated to a private ownership economy, the concept of β-quasiequilibrium fits with the quasiequilibrium notion of this economy as defined in the beginning of Chapter 2.

Definition 3.1.3 differs from an alternative definition given by Greenberg [99], related to Debreu's [59]. If correspondences β_i are assumed to be nonempty valued, a β-quasiequilibrium is what is called an equilibrium in Borglin–Keiding [41] and Yannelis–Prabhakar [180].

Before stating in Proposition 3.1.1 sufficient conditions for the existence of a β-quasiequilibrium of an abstract economy, we recall the fixed point result (See Chapter 1, Part. 3 of Proposition 1.5.5) that we will use in its proof.

LEMMA 3.1.1 (GALE–MAS-COLELL) *Given $X = \prod_{i=1}^{m} X_i$ where each X_i is a nonempty compact convex subset of some finite dimensional Euclidean vector space, let for each i, $\varphi_i : X \to X_i$ be a convex (possibly empty) valued correspondence. Assume that for every i, φ_i is lower semicontinuous. Then there exists $\overline{x} \in X$ such that for each $i = 1, \ldots, m$,*

$$\textit{either } \varphi_i(\overline{x}) = \emptyset \ \textit{ or } \overline{x}_i \in \varphi_i(\overline{x}).$$

The next proposition is one of the main results of this chapter.

PROPOSITION 3.1.1 *Let $\Gamma = ((X_i, \alpha_i, P_i)_{i \in N})$ be an abstract economy and for each $i \in N$, $\beta_i : \prod_{k \in N} X_k \to X_i$ satisfying (3.1) and (3.2). Assume that for every $i \in N$,*

1 X_i is a nonempty compact convex subset of some finite dimensional Euclidean vector space,

2 α_i is an upper semicontinuous and nonempty closed convex valued correspondence,

3 β_i is convex valued,

4 the correspondence $x \mapsto \beta_i(x) \cap P_i(x)$ is lower semicontinuous.

Then there exists $\overline{x} \in \mathcal{A}(E)$ such that for each i,

$$\text{either } \beta_i(\overline{x}) \cap \operatorname{co} P_i(\overline{x}) = \emptyset \text{ or } \overline{x}_i \in \beta_i(\overline{x}) \cap \operatorname{co} P_i(\overline{x}).$$

Assume, moreover, that for every $i \in N$ and for each $x \in \mathcal{A}(E)$, $x_i \notin \operatorname{co} P_i(x)$. Then $\overline{x}$ is a β-quasiequilibrium of Γ. It is an equilibrium (as defined in Definition 3.1.2) provided that for every i, $\beta_i(\overline{x}) \neq \emptyset$ and $P_i(\overline{x})$ is open in X_i.

Proof. Let us define for each i, the set $G_i := \{x \in X : x_i \notin \alpha_i(x)\}$. For each $x \in G_i$, since α_i is upper semicontinuous and nonempty closed convex valued, there exist an open neighborhood U_x of x in X and an open convex subset V_x of X_i such that for all $x' \in U_x$, $x'_i \notin V_x$ and $\alpha_i(x') \subset V_x$. In particular, G_i is open (in X) and the family $(U_x)_{x \in G_i}$ is an open covering of G_i. As G_i is metrizable, thus paracompact, the family $(U_x)_{x \in G_i}$ has a locally finite closed refinement[2] $(W_j)_{j \in J}$, that is:

- $(W_j)_{j \in J}$ is a closed covering of G_i and every x of G_i has a neighborhood intersecting only finitely many W_j;
- there exists a mapping $\pi : J \to G_i$ such that for every $j \in J$, $W_j \subset U_{\pi(j)}$.

For each i, the correspondence $\delta_i : G_i \to X_i$ defined by

$$\delta_i(x) = \bigcap_{\{j \,:\, x \in W_j\}} V_{\pi(j)}$$

satisfies the following properties:

- The graph of δ_i is open in $G_i \times X_i$. Indeed, if $z_i \in \delta_i(x)$ and since the closed covering $(W_j)_{j \in J}$ is locally finite, then there exists a neighborhood W of x such that for every $x' \in W$, $\delta_i(x) \subset \delta_i(x')$. As, by construction, δ_i is open valued, this shows that the graph of δ_i is open.
- For every $x \in G_i$, $\delta_i(x)$ is convex and contains $\alpha_i(x)$. Indeed, each $V_{\pi(j)}$ is convex; on the other hand, $x \in W_j$ implies $\alpha_i(x) \subset V_{\pi(j)}$.

[2] See Subsection 1.5.1 in Chapter 1.

- For every $x \in G_i$, $x_i \notin \delta_i(x)$. Indeed, $x \in W_j \Longrightarrow x \in U_{\pi(j)} \Longrightarrow x_i \notin V_{\pi(j)}$.

In order to apply Lemma 3.1.1, let us now define for each i the "reaction" correspondence $\varphi_i : X \to X_i$ by

$$\varphi_i(x) = \begin{cases} \delta_i(x) & \text{if } x_i \notin \alpha_i(x), \\ \beta_i(x) \cap \operatorname{co} P_i(x) & \text{if } x_i \in \alpha_i(x). \end{cases}$$

It is easily verified that the correspondences φ_i satisfy the conditions of Lemma 3.1.1. Let $\overline{x} \in X$ satisfy its conclusion. Recalling that, by construction of the δ_i, $\delta_i(\overline{x}) \neq \emptyset$ and $\overline{x}_i \in \delta_i(\overline{x})$ cannot hold, one has as required $\overline{x} \in \mathcal{A}(E)$ with for each i,

$$\text{either } \beta_i(\overline{x}) \cap \operatorname{co} P_i(\overline{x}) = \emptyset \ \text{ or } \overline{x}_i \in \beta_i(\overline{x}) \cap \operatorname{co} P_i(\overline{x}).$$

The proof of the two last assertions of the proposition is obvious. □

3.2 Existence of nontransitive quasiequilibrium

Let now

$$\mathcal{E} = \left(\langle \mathbb{R}^\ell, \mathbb{R}^\ell \rangle, (X_i, P_i, \omega_i)_{i \in I}, (Y_j)_{j \in J}, (\theta_{i,j})_{\substack{i \in I \\ j \in J}}, Z \right)$$

be a private ownership economy as defined in the preamble of Chapter 2. Using the same notations as in Chapter 2, we posit on $\mathcal{E}$ the following assumptions:

1 For each $i \in I$,

(a) X_i is convex and closed,

(b) $P_i : \prod_{i \in I} X_i \times \prod_{j \in J} Y_j \times S \to X_i$ is lower semicontinuous,

(c) For each $(x, y, p) \in \mathcal{A}(\mathcal{E}) \times (S \cap Z^0)$, $x_i \notin \operatorname{co} P_i(x, y, p)$,

(d) $\omega_i \in X_i - \sum_{j \in J} \theta_{i,j} Y_j - Z$ and $\widehat{X}_i$ is compact,

(e) For each $(x, y) \in \mathcal{A}(\mathcal{E})$, one has $\displaystyle\bigcap_{p \in (S \cap Z^0)} P_i(x, y, p) \neq \emptyset$;

2 For each $j \in J$, Y_j is convex and closed, $\widehat{Y}_j$ is compact;

3 (a) Z is a closed convex cone contained in the negative orthant $-\mathbb{R}^\ell_+$,

(b) $Y + Z$ is closed.

The continuity assumption 1(b) weakens the corresponding one made in Chapter 2. It is also worth noticing that in 1(c), besides the irreflexivity of strict preferences (assumed for every (x, y, p)) which is constitutive of the model, the convexity assumption is only made for components of an attainable allocation.[3] Assumption 1 (e) reduces to the usual no-satiation assumption if preferences do not depend on prices.

In counterpart, in addition to the compactness of all attainable consumption sets, we now assume that all attainable production sets are compact. With preferences depending only on the consumptions of the other consumers, that is with preference correspondences $P_i : \prod_{i \in I} X_i \to X_i$, by using techniques similar to the ones used in Chapter 2, this additional assumption could be discarded.

As in Chapter 2, one can replace Assumption 2 by:

2′ Y is convex and closed, and for each j, the attainable production set $\widehat{Y}_j$ is relatively compact.

Conditions on the primitive characteristics of the economy insuring that Assumption 3 (b) holds and that attainable consumption and production sets are compact have been given in Chapter 2, Proposition 2.2.4.

3.2.1 Quasiequilibrium existence in a compact economy

In this subsection, we will restrict ourselves to private ownership economies $\mathcal{E}$ with compact consumption and production sets. In order to use the same tricks as in Chapter 2, π will denote the maximum profit function $\pi_j(p) = \sup p \cdot Y_j = \sup\{p \cdot y_j \,:\, y_j \in Y_j\}$ and $\widehat{P}_i$: $\prod_{k \in I} X_k \times \prod_{j \in J} Y_j \times S \to X_i$ the augmented preference correspondence:

$$\widehat{P}_i(x, y, p) = \{x'_i \in X_i : x'_i = x_i + \lambda(x''_i - x_i), 0 < \lambda \leq 1,\ x''_i \in P_i(x, y, p)\}.$$

A first useful proposition (free of any compactness assumption on consumption sets) precises the relation between continuity properties of P_i and $\widehat{P}_i$.

PROPOSITION 3.2.1 *Assume that X_i is convex and that P_i is lower semicontinuous on $\prod_{k \in I} X_k \times \prod_{j \in J} Y_j \times S$. The same is true for $\widehat{P}_i$ as defined above.*

[3] The reader will verify that in the transitive case, i.e. when the preferences P_i are derived from a complete preorder on X_i, an assumption as $x_i \notin \operatorname{co} P_i(x_i)$, $\forall x_i \in X_i$ is actually equivalent to the convexity of $P_i(x_i)$ for every $x_i \in X_i$.

Proof. Let $v_i \in \widehat{P}_i(x,y,p)$, $u_i \in P_i(x,y,p)$ such that $v_i = x_i + \lambda(u_i - x_i)$ for some $\lambda : 0 < \lambda \leq 1$, and $\varepsilon > 0$. In view of the lower semicontinuity of P_i, let V be a neighborhood of (x,y,p) in $\prod_{i\in I} X_i \times \prod_{j\in J} Y_j \times S$ such that $(x',y',p) \in V \Longrightarrow P_i(x',y',p') \cap B(u_i,\varepsilon) \neq \emptyset$. The set V can be chosen such that $(x',y') \in V \Longrightarrow \|x_i' - x_i\| < \varepsilon$. It now follows from the convexity of X_i that $(x',y',p') \in V \Longrightarrow \widehat{P}_i(x',y',p') \cap B(v_i,\varepsilon) \neq \emptyset$, which proves the proposition.[4] □

Let now m be the number of consumers and n the number of producers. In order to apply Proposition 3.1.1, we associate to $\mathcal{E}$ the abstract economy

$$\Gamma = ((X_i, \alpha_i, Q_i)_{i\in N})$$

and for each $i \in N$, the correspondence $\beta_i : \prod_{k\in N} X_k \to X_i$ defined as follows.

- The set of agents, $N = \{0, 1, \ldots, m, m+1, \ldots, m+n\}$, consists of the m consumers of I indexed by $i \in \{1, \ldots, m\}$, the n producers of J indexed by $i = (m+j) \in \{m+1, \ldots, m+n\}$ and an additional agent $i = 0$ whose function is to react with prices to a given excess total demand.

- For each $i \in \{1, \ldots, m\}$, X_i is the consumption set of i. we set

$$\alpha_i(x,y,p) = \{x_i \in X_i \,:\, p\cdot x_i \leq p\cdot\omega_i + \sum_{j\in J}\theta_{i,j}\pi_j(p) + \frac{1-\|p\|}{m}\},$$

$$\beta_i(x,y,p) = \{x_i \in X_i \,:\, p\cdot x_i < p\cdot\omega_i + \sum_{j\in J}\theta_{i,j}\pi_j(p) + \frac{1-\|p\|}{m}\},$$

$$Q_i(x,y,p) = \begin{cases} \widehat{P}_i(x,y,\frac{p}{\|p\|}) & \text{if } p \neq 0 \\ \bigcap_{q\in S\cap Z^0} \widehat{P}_i(x,y,q) & \text{if } p = 0 \end{cases}$$

- For each $i \in \{m+1, \ldots, m+n\}$, we simply set $X_{m+j} = \alpha_{m+j}(x,y,p) = \beta_{m+j}(x,y,p) = Y_j$ and

$$Q_{m+j}(x,y,p) = \{y_j' \in Y_j \,:\, p\cdot y_j' > p\cdot y_j\}.$$

[4] As noticed in Gale–Mas-Colell [96], if P_i has open lower sections, it does not necessarily hold that $\widehat{P}_i$ has also open lower sections. This remark is the reason of the correction by [96] of [94].

- Finally, $X_0 = \overline{B} \cap Z^0$, $\alpha_0(x,y,p) = \beta_0(x,y,p) = \overline{B} \cap Z^0$ and

$$Q_0(x,y,p) = \{p' \in \overline{B} \cap Z^0 : (p'-p) \cdot (\sum_{i \in I} x_i - \omega - \sum_{j \in J} y_j) > 0\}.$$

We next prove the following proposition.

PROPOSITION 3.2.2 *Assume in addition to Assumptions 1 (a)–(e), 2 and 3 on $\mathcal{E}$ that all X_i and Y_j are compact. Then $\mathcal{E}$ has a quasiequilibrium.*

Proof. For the just defined abstract economy Γ, Assumption 4 of Proposition 3.1.1 for agents $i \in \{1,\dots,m\}$, i.e. for consumers, is the only one which does not immediately follow from the definitions.

We first claim that if $i \in \{1,\dots,m\}$, the lower semicontinuity of P_i implies that Q_i is lower semicontinuous. Indeed, if $p \neq 0$, the lower semicontinuity at (x,y,p) of Q_i follows easily of Proposition 3.2.1. If $p = 0$, let V be an open set in X_i such that $V \cap (\bigcap_{q \in S \cap Z^0} \widehat{P}_i(x,y,q)) \neq \emptyset$. In view of Proposition 3.2.1, for each $q \in S \cap Z^0$, there exists a neighborhood W_q of q in $S \cap Z^0$ and a neighborhood U_q of (x,y) in $\prod_{i \in I} X_i \times \prod_{j \in J} Y_j$ such that $q' \in W_q$ and $(x',y') \in U_q$ imply $V \cap \widehat{P}_i(x',y',q') \neq \emptyset$. In view of the compactness of $S \cap Z^0$, assume that $S \cap Z^0 \subset \cup_{s=1}^{r} W_{q_s}$ for some positive integer r. It follows that

$$q' \in \bigcup_{s=1}^{r} W_{q_s} \text{ and } (x',y') \in \bigcap_{s=1}^{r} U_{q_s} \Longrightarrow V \cap \widehat{P}_i(x',y',q').$$

Let $U := \bigcap_{s=1}^{r} U_{q_s}$. The set $U \subset \prod_{i \in I} X_i \times \prod_{j \in J} Y_j$ is a neighborhood of (x,y) such that $p' \in (B \cap Z^0)$, $p' \neq 0$ and $(x',y') \in U$ imply

$$V \cap \widehat{P}_i(x',y',\frac{p'}{\|p'\|}) = V \cap Q_i(x',y',p') \neq \emptyset,$$

which proves the lower semicontinuity of Q_i at $(x,y,0)$.

Finally, being the intersection of a lower semicontinuous correspondence with an open graph correspondence, the correspondence $(x,y,p) \mapsto (Q_i(x,y,p) \cap \beta_i(x,y,p))$ is lower semicontinuous.

Applying Proposition 3.1.1, let $(\overline{x},\overline{y},\overline{p}) \in \prod_{i \in I} X_i \times \prod_{j \in J} Y_j \times (\overline{B} \cap Z^0)$ be a β-quasiequilibrium of Γ. We thus have $(\overline{x},\overline{y},\overline{p}) \in \mathcal{A}(\Gamma)$ and:

$$\forall i \in I, \quad \begin{array}{ll} \text{either} & \beta_i(\overline{x},\overline{y},\overline{p}) \cap \operatorname{co} Q_i(\overline{x},\overline{y},\overline{p}) = \emptyset \\ \text{or} & \overline{x}_i \in \beta_i(\overline{x},\overline{y},\overline{p}) \cap \operatorname{co} Q_i(\overline{x},\overline{y},\overline{p}) \end{array} \tag{3.3}$$

$$\forall j \in J,\ \forall y_j \in Y_j,\ \overline{p} \cdot y_j \leq \overline{p} \cdot \overline{y}_j = \pi_j(\overline{p}), \tag{3.4}$$

$$\forall p \in \overline{B} \cap Z^0,\ p \cdot (\sum_{i \in I} \overline{x}_i - \omega - \sum_{j \in J} \overline{y}_j) \leq \overline{p} \cdot (\sum_{i \in I} \overline{x}_i - \omega - \sum_{j \in J} \overline{y}_j). \tag{3.5}$$

Using (3.5), we first prove that $(\overline{x}, \overline{y}) \in \mathcal{A}(\mathcal{E})$. Indeed, if $\overline{p} \cdot (\sum_{i \in I} \overline{x}_i - \omega - \sum_{j \in J} \overline{y}_j) \neq 0$, it follows from (3.5) that $\overline{p} \cdot (\sum_{i \in I} \overline{x}_i - \omega - \sum_{j \in J} \overline{y}_j) > 0$, thus that $\|\overline{p}\| = 1$. Summing over $i \in I$, in this case, the relations expressing that $\overline{x}_i \in \alpha_i(\overline{x}, \overline{y}, \overline{p})$, we get $\overline{p} \cdot (\sum_{i \in I} \overline{x}_i - \omega - \sum_{j \in J} \overline{y}_j) \leq 0$, which yields a contradiction. We thus have proved that $\overline{p} \cdot (\sum_{i \in I} \overline{x}_i - \omega - \sum_{j \in J} \overline{y}_j) = 0$. Consequently, $\forall p \in \overline{B} \cap Z^0$, $p \cdot (\sum_{i \in I} \overline{x}_i - \omega - \sum_{j \in J} \overline{y}_j) \leq 0$ and $(\sum_{i \in I} \overline{x}_i - \omega - \sum_{j \in J} \overline{y}_j) \in Z$, i.e. $(\overline{x}, \overline{y}) \in \mathcal{A}(\mathcal{E})$.

We next prove that $\|\overline{p}\| = 1$, that is $\overline{p} \in S$. Indeed, since $(\overline{x}, \overline{y}) \in \mathcal{A}(\mathcal{E})$, it easily follows from Assumption 1 (c) and the definition of Q_i that for every $i \in I$, $\overline{x}_i \notin \operatorname{co} Q_i(\overline{x}, \overline{y}, \overline{p})$. It thus follows from (3.3) that for every $i \in I$, $\beta_i(\overline{x}, \overline{y}, \overline{p}) \cap \operatorname{co} Q_i(\overline{x}, \overline{y}, \overline{p}) = \emptyset$, and, in view of Assumption 1 (e), the definition of Q_i and (3.4), $\overline{p} \cdot \overline{x}_i = \overline{p} \cdot \omega_i + \sum_{j \in J} \theta_{i,j} \pi_j(\overline{p}) + \frac{1 - \|\overline{p}\|}{m} = \overline{p} \cdot \omega_i + \sum_{j \in J} \theta_{i,j} \overline{p} \cdot \overline{y}_j + \frac{1 - \|\overline{p}\|}{m}$. Summing over $i \in I$, one gets $0 = 1 - \|\overline{p}\|$, that is $\overline{p} \in S$.

We now have for every $i \in I$, $\overline{p} \cdot \overline{x}_i = \overline{p} \cdot \omega_i + \sum_{j \in J} \theta_{i,j} \overline{p} \cdot \overline{y}_j$ and

$$x'_i \in \widehat{P}_i(\overline{x}, \overline{y}, \overline{p}) \Longrightarrow \overline{p} \cdot x'_i \geq \overline{p} \cdot \omega_i + \sum_{j \in J} \theta_{i,j} \overline{p} \cdot \overline{y}_j.$$

With (3.4), $(\overline{x}, \overline{y}) \in \mathcal{A}(\mathcal{E})$ and $\overline{p} \in S \cap Z^0$, the last relations prove that $(\overline{x}, \overline{y}, \overline{p})$ is a quasiequilibrium of $\mathcal{E}$. □

3.2.2 Quasiequilibrium existence in the original economy

In this subsection, $\widehat{P}_i : (\prod_{i \in I} X_i \times \prod_{j \in J} Y_j \times S) \to X_i$ denotes for each i the augmented preference correspondence defined in the original economy,

$$\widehat{P}_i(x, y, p) = \{x'_i \in X_i : x'_i = x_i + \lambda(x''_i - x_i), 0 < \lambda \leq 1,\ x''_i \in P_i(x, y, p)\}.$$

Let us fix $\underline{x}_i \in X_i$, and $\underline{y}_{i,j} \in Y_j$ such that, as in Assumption 1 (d), $\underline{x}_i - \omega_i - \sum_{j \in J} \theta_{i,j} \underline{y}_{i,j} \in Z$ for every $i \in I$. Using the compactness of all attainable consumption and production sets (Assumption 1 (d) and 2), we can fix a positive real number r such that the closed ball $\overline{B}(0,r)$ with center 0 and radius r contains in its interior all $\widehat{X}_i$, $\widehat{Y}_j$ and $\underline{x}_i$, $\underline{y}_{i,j}$. In the proof of the next proposition, we will apply Proposition 3.2.2 to the compact economy

$$\mathcal{E}^r = \left(\langle \mathbb{R}^\ell, \mathbb{R}^\ell \rangle, (X_i^r, \widehat{P}_i^r, \omega_i)_{i \in I}, (Y_j^r)_{j \in J}, (\theta_{i,j})_{\substack{i \in I \\ j \in J}}, Z \right)$$

where $X_i^r = X_i \cap \overline{B}(0,r)$, $Y_j^r = Y_j \cap \overline{B}(0,r)$ and for each $(x,y,p) \in \prod_{i \in I} X_i^r \times \prod_{j \in J} Y_j^r \times (S \cap Z^0)$, $\widehat{P}_i^r(x,y,p) = \widehat{P}_i(x,y,p) \cap \overline{B}(0,r)$.

A cautious reader should notice that the just defined correspondences $\widehat{P}_i^r : (\prod_{i \in I} X_i^r \times \prod_{j \in J} Y_j^r \times (S \cap Z^0)) \to X_i^r$ do not coincide with the Gale–Mas-Colell augmented preference correspondences in the compact economy $\mathcal{E}^r$.

PROPOSITION 3.2.3 *Under the assumptions 1, 2, 3 stated above on the economy $\mathcal{E}$, there exist $(\overline{x}, \overline{y}, \overline{p}) \in \prod_{i \in I} X_i \times \prod_{j \in J} Y_j \times (S \cap Z^0)$ such that*

1 $\sum_{i \in I} \overline{x}_i - \sum \overline{y}_j - \omega \in Z;$

2 For each $i \in I$, $\overline{p} \cdot \overline{x}_i = \overline{p} \cdot \omega_i + \sum_{j \in J} \theta_{i,j} \overline{p} \cdot \overline{y}_j$ *and* $x_i \in \widehat{P}_i(\overline{x}, \overline{y}, \overline{p}) \Longrightarrow \overline{p} \cdot x_i \geq \overline{p} \cdot \overline{x}_i;$

3 For each $j \in J$, *for every* $y_j \in Y_j$, $\overline{p} \cdot y_j \leq \overline{p} \cdot \overline{y}_j$.

4 $\overline{p} \cdot (\sum_{i \in I} \overline{x}_i - \sum \overline{y}_j - \omega) = 0.$

Equivalently, the economy $\mathcal{E}$ has a quasiequilibrium.

Proof. We first verify that the economy $\mathcal{E}^r$ satisfies the assumptions of Proposition 3.2.2. Assumptions 1 (a), (d), (e), 2 and 3 are satisfied by construction of the economy $\mathcal{E}^r$. Recall (Proposition 3.2.1) that the lower semicontinuity of the correspondences P_i implies the lower semicontinuity of correspondences $\widehat{P}_i$. In order to prove the lower semicontinuity of correspondences $\widehat{P}_i^r$, let V be an open set such that $V \cap \widehat{P}_i^r(x,y,p) = V \cap \overline{B}(0,r) \cap \widehat{P}_i(x,y,p) \neq \emptyset$. It follows from the construction of $\widehat{P}_i$ that $V \cap B(0,r) \cap \widehat{P}_i(x,y,p) \neq \emptyset$. Using the lower semicontinuity of $\widehat{P}_i$, there exists a neighborhood U of (x,y,p) such that

$$(x', y', p') \in U \Longrightarrow V \cap B(0,r) \cap \widehat{P}_i(x', y', p') \neq \emptyset,$$

which implies that $V \cap \overline{B}(0,r) \cap \widehat{P}_i(x', y', p') \neq \emptyset$. Being the restriction to $\prod_{i\in I} X_i^k \times \prod_{j\in J} Y_j^k \times (S \cap Z^0)$ of a lower semicontinuous correspondence, P_i^r is lower semicontinuous. We have also already noticed that Assumption 1 (c) for $\mathcal{E}$ implies the same assumption for the economy deduced from $\mathcal{E}$ by replacing by $\widehat{P}_i$ the correspondences P_i. A fortiori, Assumption 1 (c) for $\mathcal{E}^r$ follows from Assumption 1 (c) for $\mathcal{E}$.

Applying Proposition 3.2.2, let now $(\overline{x}, \overline{y}, \overline{p}) \in \prod_{i\in I} X_i^r \times \prod_{j\in J} Y_j^r \times (S \cap Z^0)$ be a quasiequilibrium of $\mathcal{E}^r$. We then have:

$$\forall i \in I,\ \overline{p} \cdot \overline{x}_i = \overline{p} \cdot \omega_i + \sum_{j\in J} \theta_{i,j} \overline{p} \cdot \overline{y}_j\ , \tag{3.6}$$

$$\text{and}\quad x_i' \in \widehat{P}_i^r(\overline{x}, \overline{y}, \overline{p}) \Longrightarrow \overline{p} \cdot x_i' \geq \overline{p} \cdot \overline{x}_i\ ; \tag{3.7}$$

$$\forall j \in J,\ \forall y_j \in Y_j^r,\ \overline{p} \cdot y_j \leq \overline{p} \cdot \overline{y}_j\ ; \tag{3.8}$$

$$\sum_{i\in I} \overline{x}_i - \omega - \sum_{j\in J} \overline{y}_j \in Z. \tag{3.9}$$

We now show that $(\overline{x}, \overline{y}, \overline{p})$ is also a quasiequilibrium of $\mathcal{E}$.

It follows from (3.9) that $(\overline{x}, \overline{y}) \in \mathcal{A}(\mathcal{E})$, thus that each $\overline{x}_i$ and each $\overline{y}_j$ belong to the open ball $B(0,r)$. If for some i and for some $x_i \in X_i$, $x_i \in \widehat{P}_i(\overline{x}, \overline{y}, \overline{p})$, then, in view of the convexity of X_i, there exists $\lambda : 0 < \lambda \leq 1$ such that $\overline{x}_i + \lambda(x_i - \overline{x}_i) \in \widehat{P}_i^r(\overline{x}, \overline{y}, \overline{p})$, thus that $\overline{p} \cdot (\overline{x}_i + \lambda(x_i - \overline{x}_i)) \geq \overline{p} \cdot \overline{x}_i$, that is $\overline{p} \cdot x_i \geq \overline{p} \cdot \overline{x}_i$. In the same way, for each j, let $y_j \in Y_j$ and, in view of the convexity of Y_j, let $\lambda : 0 < \lambda \leq 1$ be such that $\overline{y}_j + \lambda(y_j - \overline{y}_j) \in Y_j^r$. Then $\overline{p} \cdot (\overline{y}_j + \lambda(y_j - \overline{y}_j)) \leq \overline{p} \cdot \overline{y}_j$, that is $\overline{p} \cdot y_j \leq \overline{p} \cdot \overline{y}_j$. □

3.3 From quasiequilibrium to equilibrium

For this, we refer the reader to the thorough study done in the last section of Chapter 2.

Recall (see Proposition 2.2.5) that every quasiequilibrium of $\mathcal{E}$, in particular the quasiequilibrium proved to exist in Proposition 3.2.3, is nontrivial provided that

- $\omega \in \operatorname{int}(\sum_{i\in I} X_i - Y - Z)$.

Sufficient conditions to make a nontrivial quasiequilibrium of $\mathcal{E}$ to be an equilibrium are given in Proposition 2.3.3 and its corollaries.

Chapter 4

OPTIMALITY PROPERTIES OF EQUILIBRIUM

This chapter is devoted to the relations between the price system and the achievement of optimal economic welfare. Based on some easy consequences of the convexity assumptions made on the economy, this study was renewed in the fifties under the name of "new welfare economics".

Loosely speaking, welfare economics consist of three theoretical assertions:

- An equilibrium of a private ownership economy is optimal in several senses of the term that we shall precise later. This quasi-tautological assertion is termed "first welfare theorem".
- Social optima, in all the senses of a definition going back to Pareto, are multiple. The Paretian study of optimality considers the problem of efficient organization of an economy with an unspecified distribution of resources and profit shares to the individual consumers. The "second welfare theorem" asserts that, given an optimal attainable assignment of commodities to consumers, there is some price system to which consumers and producers are adapted. Actually, the optimal allocation and this price form an equilibrium of the private ownership economy deduced from the first one by an appropriate definition of initial resources and profit shares of consumers. The responsibility of a benevolent planner can thus be limited to the choice between different social optima and to implementing the distribution of resources and profit shares associated with the chosen optimum.
- In an economy with a specified distribution of resources and profit shares, that is, a private ownership economy, the requirement that

no coalition of consumers can make its members better off by using its own resources appears as a kind of superior form of optimality for an allocation. Obviously, an equilibrium allocation satisfies this requirement. Conversely, the "limit core equivalence theorem" asserts that if the number of consumers tends to infinity then for an allocation that no coalition can block there is at the limit a price to which consumers and producers are adapted. Actually, the allocation and this price form an equilibrium of the economy. Debreu and Scarf [63] have given a formal content to this idea originating from Edgeworth.

The three assertions are made precise in this chapter. The private ownership economy we consider

$$\mathcal{E} = \left(\langle \mathbb{R}^\ell, \mathbb{R}^\ell \rangle, (X_i, P_i, \omega_i)_{i \in I}, (Y_j)_{j \in J}, (\theta_{i,j})_{\substack{i \in I \\ j \in J}}, Z \right)$$

is one with for each i a preference correspondence $P_i : \prod_{k \in I} X_k \to X_i$ depending only on the consumptions of the other consumers and not as in the previous chapter also on prices and the current production vector. As usual, under the condition that $x_i \notin P_i(x)$, the set $P_i(x)$ is interpreted as the set of consumption vectors $x'_i \in X_i$ strictly preferred by i to x_i when the consumption vectors of all consumers $k \neq i$ are x_k. In other words, it is assumed that consumers have preferences over their consumption set with regard to the actions of other consumers. The other characteristics of the economy are the same as in Chapters 2 and 3. In particular, the disposal cone Z is a convex cone with vertex $0 \in Z$ contained in the negative orthant of $\mathbb{R}^\ell$. This assumption will be constantly used without special mention in the proofs.

Section 4.1 defines the different Pareto optimality concepts, and core, Edgeworth, fuzzy core allocations. Defining new concepts, it is in order to verify their consistency. Several nonemptiness theorems give in Section 4.2 conditions for existence of weak Pareto optimal, core, Edgeworth, fuzzy core allocations. It is worth noticing that the nonsatiation assumptions used in the equilibrium existence proofs of Chapter 1 and 2 are not needed in these nonemptiness theorems. The second welfare theorem and the limit core equivalence theorem are stated and proved in Section 4.3. Beyond the economic meaning of supporting with prices an optimal allocation, the second welfare theorem was proved to be an essential tool in the Negishi approach to the equilibrium existence problem. The same is true for the limit core equivalence theorem. With the nonemptiness result for the fuzzy core, it yields a fourth approach to the equilibrium existence problem of which an interest is to separate the fixed point argument from the separation argument, two arguments generally more intricate in the other equilibrium existence proofs.

4.1 Optimality concepts

4.1.1 Pareto dominance and Pareto optimality

Recall that

- $\omega = \sum_{i \in I} \omega_i$ is the total initial endowment;
- $Y = \sum_{j \in J} Y_j$ is the total production set;
- $\mathcal{A}(\mathcal{E}) = \left\{ (x, y) \in \prod_{i \in I} X_i \times \prod_{j \in J} Y_j : \sum_{i \in I} x_i - \sum_{j \in J} y_j - \omega \in Z \right\}$ is the set of all attainable allocations;
- $\widehat{X} = \left\{ x \in \prod_{i \in I} X_i : \exists y \in Y, \sum_{i \in I} x_i - y - \omega \in Z \right\}$ is the set of all attainable consumption allocations.

Several notions of Pareto dominance between two attainable consumption allocations x' and x in $\widehat{X}$ can be defined depending on the notion of preferences which is used, depending also whether we look for weak or for strong Pareto dominance:

1 $x'_i \in P_i(x)$ for each $i \in I$.

2 $x'_i \in \operatorname{cl} P_i(x)$ for each $i \in I$ and $x'_{i_0} \in P_{i_0}(x)$ for some $i_0 \in I$.

3 $x'_i \in \widehat{P}_i(x)$ for each $i \in I$.

4 $x'_i \in \operatorname{cl} \widehat{P}_i(x)$ for each $i \in I$ and $x'_{i_0} \in \widehat{P}_{i_0}(x)$ for some $i_0 \in I$.

When each correspondence P_i describes the (strictly) preferred sets of a complete preorder $\succsim_i$ of consumer i over his consumption set X_i, relation (2) is replaced by the standard Pareto dominance:

5 $x'_i \succsim_i x_i$ for each $i \in I$ and $x'_{i_0} \succ_{i_0} x_{i_0}$ for some $i_0 \in I$.

Relation (1) is a relation of strong Pareto dominance; relations (2) and (5) are relations of weak dominance that coincide for complete preorders which have open lower sections and are strictly quasi-convex,[1] at least when one is comparing allocations with non-satiated components. Relations (3) and (4) are relations of strong and weak Pareto dominance for the augmented preference correspondences:

$$\widehat{P}_i(x) = \{x'_i \in X_i : x'_i = x_i + \lambda(x''_i - x_i), 0 < \lambda \leq 1,\ x''_i \in P_i(x)\}.$$

Obviously, strong dominance as in (1) implies strong dominance as in (3) which implies weak dominance as in (4); strong dominance as in (1) implies weak dominance as in (5) or in (2) which implies weak dominance as in (4).

Corresponding to the previous notions we have the following definition.

DEFINITION 4.1.1 *An attainable consumption allocation $x \in \widehat{X}$ is*

1 ***weakly Pareto optimal*** *if there is no $x' \in \widehat{X}$ satisfying $x'_i \in P_i(x)$ (resp. $x'_i \succ_i x_i$) for each $i \in I$,*

2 ***Pareto optimal*** *if there is no $x' \in \widehat{X}$ satisfying $x'_i \in \operatorname{cl} P_i(x)$ (resp. $x'_i \succsim_i x_i$) for each $i \in I$ and $x'_{i_0} \in P_{i_0}(x)$ (resp. $x'_{i_0} \succ_{i_0} x_{i_0}$) for some $i_0 \in I$,*

3 ***weakly Pareto optimal for augmented preferences*** *if there is no $x' \in \widehat{X}$ satisfying $x'_i \in \widehat{P}_i(x)$ for each $i \in I$,*

4 ***Pareto optimal for augmented preferences*** *if there is no $x' \in \widehat{X}$ satisfying $x'_i \in \operatorname{cl} \widehat{P}_i(x)$ for each $i \in I$ and $x'_{i_0} \in \widehat{P}_{i_0}(x)$ for some $i_0 \in I$.*

The interpretation of weak Pareto optimality is that it is impossible to make every consumer better off. As well-known, in the case of complete preference preorders, the interpretation of Pareto optimality is that it is impossible to make one consumer better off without making another one worse off. Such an interpretation is obviously appealing given the individualistic foundations of welfare economics.

4.1.2 Core of an economy

Let us begin with some terminology. We call **coalition** any nonempty subset of the set I of consumers. If $X_S = \prod_{i \in S} X_i$, a vector $x_S \in X_S$ is

[1] A preorder $\succsim$ over a convex set X is strictly quasi-convex if $z \succ x \Longrightarrow \lambda z + (1-\lambda)x \succ x$ for every $\lambda\colon 0 < \lambda \leq 1$.

said to be a **consumption assignment** for coalition S. A **preference correspondence** $P_S\colon \prod_{i\in I} X_i \to X_S$ can be defined by

$$P_S(x) = \{x_S = (x_{iS})_{i\in S} \in X_S\colon\ x_{iS} \in P_i(x) \text{ for each } \ i \in S\}.$$

In other words, $P_S(x)$ is the set of consumption assignments x_S which are unanimously preferred to x by the members of coalition S. On the other hand, in order to define the productive power of each coalition, we assume that a coalition S owns the technology set $\sum_{i\in S}\theta_{i,j}Y_j$ at its disposal in the production set Y_j of producer j. This kind of assumption, which can be found in Aliprantis–Brown–Burkinshaw [6], Hildenbrand [103], Nikaido [147], Rader [151], lies on the idea that the relative shares $\theta_{i,j}$ reflect consumers' stock holdings which represent proprietorships of production possibilities. For each coalition S we thus define a **production possibility set** $Y_S = \sum_{i\in S}\sum_{j\in J}\theta_{i,j}Y_j$. and the set

$$\widehat{X}_S = \{x_S \in X_S\colon\ \exists y \in Y_S, \sum_{i\in S} x_i - \sum_{i\in S}\omega_i - y \in Z\}$$

is the set of all **attainable consumption assignments** for coalition S.

Definition 4.1.2 *A coalition S **blocks** an attainable consumption allocation $x \in \widehat{X}$ if there is some $x_S \in \widehat{X}_S \cap P_S(x)$.*

*The **core** of the economy $\mathcal{E}$ is the set $\mathcal{C}(\mathcal{E})$ of all attainable consumption allocations which are blocked by no coalition.*

It is worth noticing that a core allocation is a weakly Pareto optimal allocation since it is not blocked by the *grand coalition* (i.e., the coalition of all consumers).

4.1.3 Replica economies and limit core concepts

Let now r be any positive integer.

Definition 4.1.3 *The r-**replica** of $\mathcal{E}$ is the economy composed of r subeconomies identical to the original one*

$$\mathcal{E}^r = \Big(\langle\mathbb{R}^\ell,\mathbb{R}^\ell\rangle, (X_{i,q}, P_{i,q}, \omega_{i,q})_{\substack{i\in I\\ q=1,\dots,r}}, (Y_{j,q'})_{\substack{j\in J\\ q'=1,\dots,r}}, (\theta_{i,q,j,q'})_{\substack{i\in I, j\in J\\ q,q'=1,\dots,r}}, Z\Big)$$

defined as follows:

- *For each $j \in J$, the r producers of type j have the same production set $Y_{j,q'} = Y_j$.*

- *For each $i \in I$, the r consumers of type i have the same consumption set $X_{i,q} = X_i$ and the same initial endowment $\omega_{i,q} = \omega_i$.*

- *For externalities in preferences and relative profit shares $\theta_{i,q,j,q'}$, each consumer (i,q) is restricted within his subeconomy. Moreover, the correspondence $P_{i,q}\colon \prod_{k\in I} X_k \to X_i$ is defined by $P_{i,q}(x) = P_i(x)$ for each $q = 1, \ldots, r$ and $\theta_{i,q,j,q'} = \begin{cases} \theta_{i,j} & \text{if } q = q' \\ 0 & \text{if } q \neq q' \end{cases}$.*

In $\mathcal{E}^r$, a consumption allocation $(x_{i,q}) \in \prod_{\substack{i\in I\\ q=1,\ldots,r}} X_{i,q}$ such that for every $i \in I$, $x_{i,q} = x_i$ for all $q = 1, \ldots, r$ is said to have the *equal treatment property*[2]. As easily verified, such an equal treatment consumption allocation is attainable in $\mathcal{E}^r$ if and only the consumption allocation $x = (x_i)_{i\in I}$ is attainable in the original economy $\mathcal{E}$. Moreover, provided that each Y_j is convex, such an equal treatment consumption allocation belongs to the core $\mathcal{C}(\mathcal{E}^r)$ of the economy $\mathcal{E}^r$ if and only if there is no coalition $S \subset I \times \{1, \ldots, r\}$ and some consumption assignment x_S such that:

$$\sum_{(i,q)\in S} x_{i,q,S} - \sum_{i\in I} |S(i)|\omega_i \in \sum_{i\in I} |S(i)| \sum_{j\in J} \theta_{i,j} Y_j + Z, \text{ and} \tag{4.1}$$

$$x_{i,q,S} \in P_i(x) \text{ for all } (i,q) \in S, \tag{4.2}$$

where $S(i) := \{q \in \{1, \ldots, r\}\colon (i,q) \in S\}$ and $|S(i)|$ denotes the number of elements of $S(i)$.

If we assume that each X_i, Y_j and every $P_i(x)$ are convex,[3] letting $t_i = \frac{|S(i)|}{r}$, $t = (t_i)_{i\in I}$, and for each i such that $t_i > 0$,

$$x_{i,t} = \frac{1}{|S(i)|} \sum_{q\in S(i)} x_{i,q,S},$$

the scalar t_i is a rational number in $[0,1]$ which can be thought of as the **rate of participation** of i in coalition S and $x_{i,t}$ is the mean consumption that consumer i achieves by participating in the coalition. Let us

[2]For an exchange economy $\mathcal{E}$ with continuous, transitive and complete, strictly convex and strictly monotone consumers' preferences, Debreu and Scarf [63] established that any core allocation of $\mathcal{E}^r$ has the equal treatment property.

[3]These convexity assumptions will be maintained in the sequel of this chapter.

replace (4.1) and (4.2) by

$$\sum_{\{i\in I:\ t_i>0\}} t_i x_{i,t} - \sum_{i\in I} t_i\omega_i \in \sum_{i\in I} t_i \sum_{j\in J} \theta_{i,j} Y_j + Z, \text{ and} \tag{4.3}$$

$$x_{i,t} \in P_i(x) \text{ for all } i \text{ such that } t_i > 0. \tag{4.4}$$

The existence of some x_S satisfying (4.1) and (4.2) is obviously equivalent to the existence of some x_t satisfying (4.3) and (4.4).

Let us denote by $\mathcal{C}^r(\mathcal{E})$ the set of all attainable consumption allocations of $\mathcal{E}$ whose r-replica belongs to $\mathcal{C}(\mathcal{E}^r)$. Clearly, $\mathcal{C}^{r+1}(\mathcal{E}) \subset \mathcal{C}^r(\mathcal{E})$. The previous considerations lead to the following two definitions.

DEFINITION 4.1.4 *An attainable consumption allocation $x \in \widehat{X}$ is an **Edgeworth equilibrium** if $x \in \bigcap_{r\geq 1} \mathcal{C}^r(\mathcal{E})$.*

That is, an attainable consumption allocation is an Edgeworth equilibrium of an economy if for any positive integer r its r-replica belongs to the core of the r-replica economy. The set of all Edgeworth equilibria is denoted by

$$\mathcal{C}^e(\mathcal{E}) := \bigcap_{r\geq 1} \mathcal{C}^r(\mathcal{E}).$$

This notion was introduced in [5]; the ideas originate from [63].

Let $\mathbb{Q}$ denote the set of rational numbers. A glance to (4.3) and (4.4) shows that an attainable consumption allocation x belongs to $\mathcal{C}^e(\mathcal{E})$ if and only if there is no $t \in ([0,1] \cap \mathbb{Q})^I$, $t \neq 0$, and $x_t = (x_{i,t}) \in \prod_{\{i\in I:\ t_i>0\}} X_i$ satisfying the conditions (4.3) and (4.4). Allowing as Aubin [23] the rates of participation t_i to take any value in the real interval $[0,1]$, a nonzero vector $t = (t_i) \in [0,1]^I$ is called **fuzzy coalition**. With this definition, a coalition S, that is a nonempty subset S of I, can be identified with the fuzzy coalition $t = (t_i) \in [0,1]^I$ such that

$$t_i = \begin{cases} 1 & \text{if } i \in S; \\ 0 & \text{otherwise,} \end{cases}$$

and we have the following definition.

DEFINITION 4.1.5 *A fuzzy coalition t **blocks** an attainable consumption allocation $x \in \widehat{X}$ if there is some $x_t = (x_{i,t}) \in \prod_{\{i:\ t_i>0\}} X_i$ such that*

$$\sum_{i\in I} t_i x_{i,t} - \sum_{i\in I} t_i\omega_i \in \sum_{i\in I} t_i \sum_{j\in J} \theta_{i,j} Y_j + Z, \text{ and} \tag{4.5}$$

$$x_{i,t} \in P_i(x) \text{ for each } i \text{ such that } t_i > 0. \tag{4.6}$$

*The **fuzzy core** of the economy $\mathcal{E}$ is the set $\mathcal{C}^f(\mathcal{E})$ of all attainable consumption allocations which are blocked by no fuzzy coalition.*

Let $\mathcal{T} = [0,1]^I \setminus \{0\}$ be the set of all fuzzy coalitions. For each $t \in \mathcal{T}$, define

$$X_t := \prod_{\{i:\ t_i>0\}} X_i; \quad Y_t := \sum_{i\in I} t_i \sum_{j\in J} \theta_{i,j} Y_j;$$

$$\widehat{X}_t := \Big\{ x_t \in X_t \colon \sum_{i\in I} t_i x_{i,t} - \sum_{i\in I} t_i \omega_i \in Y_t + Z \Big\}.$$

The sets Y_t and $\widehat{X}_t$ can be respectively thought of as the **production possibility set** and the **set of all attainable consumption assignments** for the fuzzy coalition t. On the other hand, if $x \in \prod_{i\in I} X_i$,

$$P_t(x) := \{x_t \in X_t \colon\ x_{i,t} \in P_i(x) \text{ for each } i \text{ such that } t_i > 0\}$$

defines a **preference correspondence** $P_t \colon \prod_{i\in I} X_i \to X_t$ for the fuzzy coalition t.

The notion of blocking an attainable consumption allocation $x \in \widehat{X}$ for a fuzzy coalition t and the notion of fuzzy core can now be rephrased as:

*The fuzzy coalition t **blocks** the attainable consumption allocation $x \in \widehat{X}$ if there is some $x_t \in \widehat{X}_t \cap P_t(x)$. The **fuzzy core** of $\mathcal{E}$ is the set of all attainable consumption allocations such that $\widehat{X}_t \cap P_t(x) = \emptyset$ holds for every $t \in \mathcal{T}$.*

Clearly, if we assume that each X_i, Y_j and $P_i(x)$ for every $x \in \widehat{X}$ are convex then

$$\mathcal{C}^f(\mathcal{E}) \subset \mathcal{C}^e(\mathcal{E}) = \bigcap_{r\geq 1} \mathcal{C}^r(\mathcal{E}) \subset \mathcal{C}(\mathcal{E}).$$

4.2 Nonemptiness theorems

We begin with the existence of optimal allocations. The existence of weakly Pareto optimal consumption allocations when preference correspondences have convex values and open lower sections is an easy consequence of an elementary fixed point theorem.

PROPOSITION 4.2.1 *The economy $\mathcal{E}$ has weakly Pareto optimal allocations provided the following conditions are satisfied:*

1. *The set $\widehat{X}$ of all attainable consumption allocations is nonempty convex and compact;*

2. *For each i the correspondence $P_i: \prod_{k \in I} X_k \to X_i$ is convex valued;*

3. *For every $z \in \widehat{X}$ and for each i, the set $P_i^{-1}(z_i)$ is open in X.*

Proof. Let us define the correspondence $P: \widehat{X} \to \widehat{X}$ by $P(x) = \{x' \in \widehat{X}: x'_i \in P_i(x) \text{ for each } i \in I\}$. It is readily verified that P has open lower sections and convex values with $x \notin P(x)$. The existence of $x \in \widehat{X}$ such that $P(x) = \emptyset$ follows from the Browder–Fan theorem (Proposition 1.2.1). To conclude, notice that weakly Pareto optimal consumption allocations are precisely the elements x of $\widehat{X}$ such that $P(x) = \emptyset$. □

Proposition 2.2.4 in Chapter 2 give sufficient conditions on the primitive characteristics of the economy insuring that the convexity and compactness of $\widehat{X}$ in Condition 1 above are satisfied. More precisely, the reader will verify that $\widehat{X}$ is convex and compact if the individual consumption sets, the total production set and the disposal cone are convex and closed, and if

a. $\mathrm{A}(Y) \cap (-Z) = \{0\}$,

b. $\mathrm{A}(\sum_{i \in I} X_i) \cap \mathrm{A}(Y + Z) = \{0\}$,

c. $\mathrm{A}(\sum_{i \in I} X_i) \cap (-\mathrm{A}(\sum_{i \in I} X_i)) = \{0\}$.

It is worth noticing[4] that in Proposition 4.2.1 Condition 2 is not needed if each P_i corresponds to an irreflexive and transitive binary relation on X_i.

PROPOSITION 4.2.2 *If each binary relation P_i is irreflexive and transitive, the economy $\mathcal{E}$ has weakly Pareto optimal consumption allocations provided that:*

1. *The set $\widehat{X}$ of all attainable consumption allocations is nonempty and compact;*

2. *For every $z \in \widehat{X}$ and for each i, the set $P_i^{-1}(z_i)$ is open in X_i.*

Proof. Let us define the correspondence $P: \widehat{X} \to \widehat{X}$ by $P(x) = \{x' \in \widehat{X}: x'_i \in P_i(x_i) \text{ for each } i \in I\}$. The binary relation P is irreflexive,

[4]The remark is due to Schmeidler [163].

transitive and has open lower sections in $\widehat{X}$. Assume by contraposition that $P(x) \neq \emptyset$ for every $x \in X$. Then $\widehat{X} = \bigcup_{x' \in \hat{X}} P^{-1}(x')$, thus the sets $(P^{-1}(x'))_{x' \in \hat{X}}$ form an open covering of $\widehat{X}$. Since $\widehat{X}$ is compact, let F be a finite subset of $\widehat{X}$ such that

$$\widehat{X} = \bigcup_{x' \in F} P^{-1}(x'). \tag{4.7}$$

Since F is finite, let x^* be a maximal element of F, that is, $P(x^*) \cap F = \emptyset$. In view of (4.7), there is some $x' \in F$ such that $x' \in P(x^*)$ contrary to $P(x^*) \cap F = \emptyset$. With this contradiction, the proof is complete. □

Finally, standard Pareto optimal consumption allocations exist under the same conditions when each correspondence P_i describes the (strictly) preferred sets of a complete preorder $\succsim_i$ of consumer i on his consumption set X_i.

PROPOSITION 4.2.3 *Assume that each P_i corresponds to a complete preorder $\succsim_i$ on X_i. Then the economy $\mathcal{E}$ has standard Pareto optimal consumption allocations provided that:*

1 *The set $\widehat{X}$ of all attainable consumption allocations is nonempty and compact;*

2 *For every $z \in \widehat{X}$ and for each i, the set $P_i^{-1}(z_i)$ is open in X_i.*

Proof. Recall that for each i there exists an upper semicontinuous utility function $u_i: X_i \to \mathbb{R}^\ell$ representing the preorder $\succsim_i$. Let $(\lambda_i)_{i \in I}$ be any family of strictly positive weights for each consumer. The function $u: \prod_{i \in I} X_i \to \mathbb{R}$ defined by $u(x) = \sum_{i \in I} \lambda_i u_i(x_i)$ is upper semicontinuous on the compact set $\widehat{X}$, thus has a maximum $\overline{x} = (\overline{x}_i)_{i \in I}$. Clearly this maximum element is a standard Pareto optimal consumption allocation. □

The next results establish the nonemptiness of the core and the existence of Edgeworth equilibria. We will prove these results under the following assumptions on $\mathcal{E}$:

1 The set $\widehat{X}$ is compact. Moreover, for each $i \in I$,

   (a) X_i is convex,

   (b) For each $x_i \in X_i$, $(P_i)^{-1}(x_i)$ is open in $\prod_{k \in I} X_k$,

   (c) For each $x \in \widehat{X}$, $P_i(x)$ is convex with $x_i \notin P_i(x)$,

(d) $\omega_i \in X_i - \sum_{j \in J} \theta_{i,j} Y_j - Z$;

2 For each $j \in J$, Y_j is convex;

3 Z is a convex cone contained in the negative orthant $-\mathbb{R}^\ell_+$.

The main addition to the assumptions used in Proposition 4.2.1 for getting the existence of weakly Pareto optimal allocations is that, with the autarky assumption 1.(d), we now assume that for each i the set $\widehat{X}_{\{i\}}$ of all attainable consumption assignments for the coalition $\{i\}$ (reduced to consumer i alone) is nonempty. This obviously implies that for each $t \in \mathcal{T}$ the set $\widehat{X}_t$ of all attainable consumption assignments for the coalition t is nonempty. In particular, $\widehat{X} \neq \emptyset$. In addition, due to 1.(a) and 2, each $\widehat{X}_t$ and thus $\widehat{X}$ are convex.

For any positive integer $r \geq 1$, let

$$\mathcal{T}_r := \{t = (t_i)_{i \in I} \colon\ rt_i \in \{0, 1, \ldots, r\} \text{ for each } i \in I\}.$$

Let Y' be a convex subset of $\mathbb{R}^\ell$ containing $Y + Z$ where Y is the total production set. In the next proposition, we replace $\widehat{X}$ in Assumption 1 by the convex set

$$\widehat{X}' := \{x \in X \colon \sum_{i \in I} x_i - \sum_{i \in I} \omega_i \in Y'\}.$$

In particular, we assume that $\widehat{X}'$ is compact and that $x_i \notin P_i(x)$ for each $x \in \widehat{X}'$. Let us call $1'$ this new assumption.

PROPOSITION 4.2.4 *Let r be any positive integer. Under the assumptions $1'$, 2, 3 stated above on the economy $\mathcal{E}$, there is some $\overline{x} \in \widehat{X}'$ such that $\widehat{X}_t \cap P_t(x) = \emptyset$ holds for every $t \in \mathcal{T}_r$.*

Proof. Let us define

$$\Delta_{\mathcal{T}_r} = \left\{ \lambda = (\lambda_t)_{t \in \mathcal{T}_r} \colon \begin{array}{cc} \lambda_t \geq 0 & \text{for each } t \in \mathcal{T}_r \\ \sum_{t \in \mathcal{T}_r} \lambda_t t_i = 1 & \text{for each } i \in I \end{array} \right\}.$$

The set $\Delta_{\mathcal{T}_r}$ is for the fuzzy coalitions $t \in \mathcal{T}_r$ an analogue of the associated balancing coefficients used in game theory for stating that a family of coalitions is balanced. We first observe that the economy satisfies *the balancedness property*:

$$\lambda \in \Delta_{\mathcal{T}_r} \Longrightarrow \sum_{t \in \mathcal{T}_r} \lambda_t Y_t \subset Y. \tag{4.8}$$

For each $(x, z, \lambda) \in \widehat{X}' \times \prod_{t \in \mathcal{T}_r} \operatorname{cl} \widehat{X}_t \times \Delta_{\mathcal{T}_r}$ we define

$$I(x) = \{t \in \mathcal{T}_r\colon\ \widehat{X}_t \cap P_t(x) \neq \emptyset\}$$

and the function and correspondences

- $\theta(z, \lambda) = (x'_i)_{i \in I}$ with for each $i \in I$, $x'_i = \sum_{t \in \mathcal{T}_r} \lambda_t t_i z_{i,t}$.
- $\varphi(x) = (\varphi_t(x))_{t \in \mathcal{T}_r}$ with for each $t \in \mathcal{T}_r$, $\varphi_t(x) = \operatorname{cl} \widehat{X}_t \cap P_t(x)$.
- $\psi(x, \lambda) = \begin{cases} \bigcap_{t \in I(x)} \{\mu \in \Delta_{\mathcal{T}_r}\colon\ \mu_t > \lambda_t\} & \text{if } I(x) \neq \emptyset; \\ \emptyset & \text{if } I(x) = \emptyset. \end{cases}$

We first claim that $\widehat{X}' \times \prod_{t \in \mathcal{T}_r} \operatorname{cl} \widehat{X}_t \times \Delta_{\mathcal{T}_r}$ is a nonempty convex and compact subset of some Euclidean space. Indeed, it should be clear that each component of this product is nonempty and convex. By Assumption 1′, the set $\widehat{X}'$ is compact. For each i, let $\underline{x}_i$ be an autarky consumption (by Assumption 1.(d) $\underline{x}_i \in X_i \cap (\omega_i + \sum \theta_{i,j} Y_j + Z)$). To see that each $\widehat{X}_t$ is relatively compact, let us associate with each $x_t \in \widehat{X}_t$ the consumption allocation whose component i is equal to $t_i x_{i,t} + (1 - t_i)\underline{x}_i$. This consumption allocation is in $\widehat{X}'$ which is compact. Thus $\widehat{X}_t$ is relatively compact and $\operatorname{cl} \widehat{X}_t$ is compact. Finally, given the definition of $\Delta_{\mathcal{T}_r}$, to prove its compactness is routine.

By definition, the correspondences φ_t and ψ take their values in respectively $\operatorname{cl} \widehat{X}_t$ and $\Delta_{\mathcal{T}_r}$. We now prove that the function θ takes its values in $\widehat{X}'$. Indeed, if $z = (z_{i,t}) \in \prod_{t \in \mathcal{T}_r} \widehat{X}_t$, $\lambda \in \Delta_{\mathcal{T}_r}$, $x' = \theta(z, \lambda)$, then it follows from the balancedness property (4.8) that

$$\sum_{i \in I} x'_i = \sum_{i \in I} \sum_{t \in \mathcal{T}_r} \lambda_t t_i z_{i,t} = \sum_{t \in \mathcal{T}_r} \lambda_t \sum_{i \in I} t_i z_{i,t} \in \sum_{t \in \mathcal{T}_r} \lambda_t (\sum_{i \in I} t_i \omega_i + Y_t + Z)$$

$$\subset \sum_{i \in I} (\sum_{t \in \mathcal{T}_r} \lambda_t t_i) \omega_i + Y' = \sum_{i \in I} \omega_i + Y'.$$

Since the function θ is clearly continuous and the set $\widehat{X}'$ is assumed to be compact, if $z = (z_{i,t}) \in \prod_{t \in \mathcal{T}_r} \operatorname{cl} \widehat{X}_t$ and $\lambda \in \Delta_{\mathcal{T}_r}$, then $x' = \theta(z, \lambda) \in \widehat{X}'$.

On the other hand, the function θ is continuous. By Assumption 1′.(b) each correspondence φ_t defined on $\widehat{X}'$ by $\varphi_t(x) = \operatorname{cl} \widehat{X}_t \cap P_t(x)$ has open lower sections. We now define an open graph correspondence $\psi'\colon \widehat{X}' \times \Delta_{\mathcal{T}_r} \to \Delta_{\mathcal{T}_r}$ such that $\psi(x, \lambda) \subset \psi'(x, \lambda)$ for each $(x, \lambda) \in$

$\widehat{X}' \times \Delta_{\mathcal{T}_r}$.[5] For each $x \in \widehat{X}'$ such that $I(x) \neq \emptyset$, let U_x be an open neighborhood (in $\widehat{X}'$) of x such that $I(x) \subset I(x')$ for each $x' \in U_x$. Letting

$$\psi_x(z,\lambda) := \bigcap_{t\in I(x)} \{\mu \in \Delta_{\mathcal{T}_r} \colon \mu_t > \lambda_t\},$$

we thus have $\psi(x',\lambda) \subset \psi_x(x',\lambda)$ for each $x' \in U_x$. The family $(U_x)_{x\in\widehat{X}'}$ is an open covering of $\widehat{X}'$ which, by compactness of $\widehat{X}'$, contains a finite subcovering $(U_{x_i})_{i\in J}$, J finite and has a closed refinement $(F_i)_{i\in J}$. That is, $\widehat{X}' = \bigcup_{i\in J} F_i$ with each F_i closed, $F_i \subset U_{x_i}$ for each $i \in J$. Let

$$\psi_i(x,\lambda) = \begin{cases} \psi_x(x,\lambda) & \text{if } x \in F_i; \\ \Delta_{\mathcal{T}_r} & \text{if } x \notin F_i, \end{cases}$$

and

$$\psi'(x,\lambda) = \bigcap_{i\in J} \psi_i(x,\lambda).$$

It is easily verified that each correspondence ψ_i has an open graph, that ψ' has an open graph and that, as required,

$$\psi(x,\lambda) \subset \psi'(x,\lambda) \quad \text{for each} \quad (x,\lambda) \in \widehat{X}' \times \Delta_{\mathcal{T}_r}.$$

Let $\chi \colon \widehat{X}' \times \prod_{t\in\mathcal{T}_r} \operatorname{cl}\widehat{X}_t \times \Delta_{\mathcal{T}_r} \to \widehat{X}' \times \prod_{t\in\mathcal{T}_r} \operatorname{cl}\widehat{X}_t \times \Delta_{\mathcal{T}_r}$ be the correspondence defined by

$$\chi(x,z,\lambda) = \Big(\theta(z,\lambda), (\varphi_t(x))_{t\in\mathcal{T}_r}, \psi'(x,\lambda)\Big).$$

From the Gale and Mas-Colell fixed point theorem (see Proposition 1.5.5 in Chapter 1), it follows that for some $(\overline{x},\overline{z},\overline{\lambda}) \in \widehat{X}' \times \prod_{t\in\mathcal{T}_r} \operatorname{cl}\widehat{X}_t \times \Delta_{\mathcal{T}_r}$ we have:

1 $\overline{x} = \theta(\overline{z},\overline{\lambda})$;

2 for each $t \in \mathcal{T}_r$, either $\overline{z}_t \in \operatorname{cl}\widehat{X}_t \cap P_t(\overline{x})$ or $\operatorname{cl}\widehat{X}_t \cap P_t(\overline{x}) = \emptyset$;

3 $\psi'(\overline{x},\overline{\lambda}) = \emptyset$, and thus $\psi(\overline{x},\overline{\lambda}) = \emptyset$.

To complete the proof, we show by contraposition that $I(\overline{x}) = \emptyset$. Indeed, if not, recall that for each $i \in I$ the coalition $\{i\}$ can be identified

[5] The construction made here is similar to the one used in the proof of Corollary 1.2.3 in Chapter 1.

with the ith vector of the natural basis of $\mathbb{R}^I$ which belongs to $\mathcal{T}_r$. It is then easy to deduce from 3 that for some $i_0 \in I$,

$$\sum_{t \in I(\overline{x})} \overline{\lambda}_t t_{i_0} = 1 \quad \lambda_t t_{i_0} = 0 \text{ for all } t \notin I(\overline{x}).$$

Then by 1, $\overline{x}_{i_0} = \sum_{t \in I(\overline{x})} \overline{\lambda}_t t_i \overline{z}_{i,t}$ and, by 2, $\overline{x}_{i_0} \in \operatorname{co} P_{i_0}(\overline{x}) = P_{i_0}(\overline{x})$, which contradicts Assumption 1′. □

Proposition 4.2.4 has an obvious corollary which is the version of the result to be used in the sequel of this section. The slightly more general result stated in Proposition 4.2.4 will be used in the next chapter.

COROLLARY 4.2.1 *Let r be any positive integer. Under the assumptions 1, 2, 3 stated above on the economy $\mathcal{E}$, the set $\mathcal{C}^r(\mathcal{E})$ is nonempty.*

PROPOSITION 4.2.5 *Under the assumptions 1, 2, 3 stated above, the economy $\mathcal{E}$ has Edgeworth equilibria, that is, $\mathcal{C}^e(\mathcal{E}) \neq \emptyset$.*

Proof. Recall that $\mathcal{C}^e(\mathcal{E}) = \bigcap_{r \geq 1} \mathcal{C}^r(\mathcal{E})$ and that $\mathcal{C}^{r+1}(\mathcal{E}) \subset \mathcal{C}^r(\mathcal{E})$ for every $r \geq 1$. It follows from the definition and Assumption 1.(b) that each $\mathcal{C}^r(\mathcal{E})$ is a closed subset of $\widehat{X}$, nonempty by Corollary 4.2.1. Then the nonemptiness of $\mathcal{C}^e(\mathcal{E})$ follows from the compactness of $\widehat{X}$. □

In order to prove the nonemptiness of the fuzzy core, we posit now an additional assumption:

4 For each i, for every $x \in \widehat{X}$, if $z_i \in P_i(x)$ and $v_i \in X_i$ then there exists $\lambda : 0 \leq \lambda < 1$ such that $(\lambda z_i + (1 - \lambda)v_i) \in P_i(x)$.

This condition, borrowed from Bergstrom [30], was used in Chapter 2 to pass from a quasiequilibrium to an equilibrium. As already noticed (see Remark 2.3.1), Assumption 4 is satisfied if for each i and for every $x \in \widehat{X}$ the consumption set X_i is convex and the preferred set $P_i(x)$ is open in X_i. Assumption 4 is also satisfied if for each i, the consumption set X_i is convex and P_i is defined by a concave utility function.

We will use this additional assumption and the density of $\mathbb{Q}$, the set of rational numbers, in the set $\mathbb{R}$ of real numbers to prove the last nonemptiness result of this section.

PROPOSITION 4.2.6 *Under the assumptions 1, 2, 3, 4 stated above, the economy $\mathcal{E}$ has a nonempty fuzzy core, that is, $\mathcal{C}^f(\mathcal{E}) \neq \emptyset$.*

Proof. By proposition 4.2.5, $\mathcal{C}^e(\mathcal{E}) \neq \emptyset$. Under the additional assumption 4, we now show that $\mathcal{C}^e(\mathcal{E}) \subset \mathcal{C}^f(\mathcal{E})$.

Let $x \in \mathcal{C}^e(\mathcal{E})$. If $x \notin \mathcal{C}^f(\mathcal{E})$, then there exist $t \in \mathcal{T}$ and $x_t \in X_t$ such that

$$\sum_{\{i\in I:\ t_i>0\}} t_i x_{i,t} - \sum_{i\in I} t_i\omega_i \in \sum_{i\in I} t_i \sum_{j\in J}\theta_{i,j}Y_j + Z \tag{4.9}$$

$$x_{i,t} \in P_i(x) \text{ for each } i \text{ such that } t_i > 0. \tag{4.10}$$

Using (4.10), Assumption 4 and the convexity of $P_i(x)$, we can find some $\varepsilon > 0$ such that $0 \leq 1-\varepsilon < \lambda \leq 1$ implies

$$\lambda x_{i,t} + (1-\lambda)\underline{x}_i \in P_i(x) \text{ for each } i \text{ such that } t_i > 0,$$

where as previously $\underline{x}_i \in X_i \cap (\omega_i + \sum_{j\in J}\theta_{i,j}Y_j + Z)$ is an autarky consumption (Assumption 1.(d)) for consumer i. Let $s \in ([0,1]\cap\mathbb{Q})^I$ be such that $t_i = 0 \Longrightarrow s_i = 0$ and $t_i > 0 \Longrightarrow 1-\varepsilon < \frac{t_i}{s_i} \leq 1$.

Set $x_{i,s} = \frac{t_i}{s_i}x_{i,t} + (1-\frac{t_i}{s_i})\underline{x}_i$ for each i such that $t_i > 0$, and notice that $x_{i,s} \in P_i(x_i)$. Using (4.9), a simple calculation shows that

$$\sum_{\{i\in I:\ s_i>0\}} s_i x_{i,s} - \sum_{i\in I} s_i\omega_i \in \sum_{\{i\in I:\ s_i>0\}} s_i\frac{t_i}{s_i}\sum_{j\in J}\theta_{i,j}Y_j$$

$$+ \sum_{\{i\in I:\ s_i>0\}} s_i(1-\frac{t_i}{s_i})\sum_{j\in J}\theta_{i,j}Y_j + Z \subset \sum_{i\in I} s_i \sum_{j\in J}\theta_{i,j}Y_j + Z$$

by convexity of the Y_j. This contradicts $x \in \mathcal{C}^e(\mathcal{E})$. □

4.3 Decentralization theorems

As we will see immediately, it follows from the very equilibrium definitions given in the preamble to Chapter 2 that for an economy $\mathcal{E}$ an equilibrium consumption allocation is an element of the fuzzy core, thus an Edgeworth equilibrium, an element of the core and a weakly Pareto optimal allocation. An equilibrium consumption allocation obtained in Proposition 2.3.3 is also a Pareto optimal allocation for augmented preferences.

PROPOSITION 4.3.1 *Let $(\overline{x}, \overline{y}, \overline{p})$ be an equilibrium of $\mathcal{E}$. Then $\overline{x} = (\overline{x}_i)_{i\in I} \in \mathcal{C}^f(\mathcal{E})$.*

Assume on the contrary that $\overline{x} = (\overline{x}_i)_{i\in I} \notin \mathcal{C}^f(\mathcal{E})$. Let t be a fuzzy coalition and $x_t \in X_t$ such that

$$\sum_{\{i\in I:\ t_i>0\}} t_i x_{i,t} - \sum_{i\in I} t_i\omega_i \in \sum_{i\in I} t_i \sum_{j\in J}\theta_{i,j}Y_j + Z$$

$$x_{i,t} \in P_i(\overline{x}) \text{ for each } i \text{ such that } t_i > 0.$$

From the first relation and the equilibrium definition we deduce

$$\overline{p} \cdot \Big(\sum_{\{i \in I:\ t_i > 0\}} t_i x_{i,t} - \sum_{i \in I} t_i \omega_i \Big) \leq \overline{p} \cdot \sum_{i \in I} t_i \sum_{j \in J} \theta_{i,j} \overline{y}_j. \tag{4.11}$$

From the second relation and the equilibrium definition we deduce

$$\overline{p} \cdot x_{i,t} > \overline{p} \cdot \omega_i + \sum_{j \in J} \theta_{i,j} \overline{p} \cdot \overline{y}_j \text{ for each } i \text{ such that } t_i > 0$$

and

$$\overline{p} \cdot \Big(\sum_{\{i \in I:\ t_i > 0\}} t_i x_{i,t} - \sum_{i \in I} t_i \omega_i \Big) > \overline{p} \cdot \sum_{i \in I} t_i \sum_{j \in J} \theta_{i,j} \overline{y}_j, \tag{4.12}$$

which contradicts (4.11). □

PROPOSITION 4.3.2 *Let $(\overline{x}, \overline{y}, \overline{p})$ be an equilibrium of $\mathcal{E}$ obtained in Proposition 2.3.3 from a nontrivial quasiequilibrium satisfying the conditions stated in Proposition 2.2.2 or in Proposition 3.2.3. Then $\overline{x} = (\overline{x}_i)_{i \in I}$ is a Pareto optimal consumption allocation for augmented preferences, thus a Pareto optimal consumption allocation.*

Proof. Assume on the contrary that there is some $x = (x_i)_{i \in I} \in \widehat{X}$, that is, $\sum_{i \in I}(x_i - \omega_i) \in Y + Z$, such that $x_i \in \mathrm{cl}\, \widehat{P}_i(\overline{x})$ for each $i \in I$ and $x_{i_0} \in \widehat{P}_{i_0}(\overline{x})$ for some $i_0 \in I$. From the definition of equilibrium, we get

$$\overline{p} \cdot \sum_{i \in I} x_i > \overline{p} \cdot \sum_{i \in I} \overline{x}_i = \overline{p} \cdot \sum_{i \in I} \omega_i + \sum_{i \in I} \sum_{j \in J} \theta_{i,j} \overline{p} \cdot \overline{y}_j \geq \overline{p} \cdot \sum_{i \in I} \omega_i + \overline{p} \cdot y$$

for every $y \in Y$. Recalling that, by definition of equilibrium, $\overline{p} \cdot Z \leq 0$, we have got a contradiction. □

The main purpose of this section is to get converse results.

4.3.1 Supporting weakly Pareto optimal allocations

Before stating the precise result, we introduce a new equilibrium concept for the economy $\mathcal{E}$ and its corresponding quasiequilibrium concept.

DEFINITION 4.3.1 *An attainable allocation $(\overline{x}, \overline{y})$ is said to be* ***an equilibrium relative to the price system*** *$p \in \mathbb{R}^\ell$ if:*

1 *For each* $i \in I$, $x_i \in P_i(\overline{x}) \Longrightarrow p \cdot x_i > p \cdot \overline{x}_i$,

2 *For each* $j \in J$, $y_j \in Y_j \Longrightarrow p \cdot y_j \leq p \cdot \overline{y}_j$,

3 $p \in Z^0 := \{p' \in \mathbb{R}^\ell : p' \cdot z \leq 0, \ \forall z \in Z\}$
where Z^0 *is the polar cone of* Z, *and*

$$p \cdot \sum_{i \in I} \overline{x}_i = p \cdot \sum_{j \in J} \overline{y}_j + p \cdot \sum_{i \in I} \omega_i.$$

Condition (1) means that each consumer best satisfies his preferences subject to a wealth constraint precisely equal to the value of his equilibrium consumption. Condition (2) and (3) are the classical conditions of profit maximization subject to technological constraints and minimization of the cost of the equilibrium disposal for the given price system.

DEFINITION 4.3.2 *An attainable allocation* $(\overline{x}, \overline{y})$ *is said to be* ***a quasi-equilibrium relative to the price system*** $p \in \mathbb{R}^\ell$ *if the first condition of the previous definition is replaced by:*

1′ *For each* $i \in I$, $x_i \in P_i(\overline{x}) \Longrightarrow p \cdot x_i \geq p \cdot \overline{x}_i$.

The quasiequilibrium $(\overline{x}, \overline{y})$ *is said* ***nontrivial*** *if for some* $i \in I$ *and for some* $x_i \in X_i$, $p \cdot x_i < p \cdot \overline{x}_i$.

Assume local no-satiation for each consumer at his component of $\overline{x}$. Here Condition (1′) states that for each i, the consumption $\overline{x}_i$ minimizes the expenditure of consumer i over the set $\operatorname{cl} P_i(\overline{x})$.

REMARK 4.3.1 For an attainable allocation $(\overline{x}, \overline{y})$, set

$$z = \sum_{i \in I} \overline{x}_i - \sum_{j \in J} \overline{y}_j - \omega \in Z.$$

It is easy to check that if $(\overline{x}, \overline{y})$ is an equilibrium (resp. a quasiequilibrium) relative to the price system p, then $(\overline{x}, \overline{y}, p)$ is an equilibrium (resp. a quasiequilibrium) of a private ownership economy $\mathcal{E}'$ with the same attainable consumption allocations, thus the same weakly Pareto optimal consumption allocations as $\mathcal{E}$, obtained from $\mathcal{E}$ by giving to the ith consumer the resources $\omega_i = \overline{x}_i - \frac{1}{|I|} \sum_{j \in J} \overline{y}_j - \frac{1}{|I|} z$ and the shares $\theta_{i,j} = \frac{1}{|I|}$.[6] As noticed by Debreu [58], the interest of the "relative to

[6]Recall that $|I|$ denotes the number of elements of I, that is, the number of consumers.

a price system" equilibrium concepts is that, like the Pareto optimality concepts, they do not require any specification of the resources and the shares of consumers.

The concept of quasiequilibrium relative to a price system is used in the following important result.

PROPOSITION 4.3.3 (GALE–MAS-COLELL [95]) *Let $\overline{x}$ be a weakly Pareto consumption allocation of $\mathcal{E}$ and let $\overline{y} \in \prod_{j\in J} Y_j$ be such that $\sum_{i\in I} \overline{x}_i - \omega - \sum_{j\in J} \overline{y}_j = z \in Z$. There is a nonzero price system p such that $(\overline{x}, \overline{y})$ is a quasiequilibrium relative to p, provided the following conditions are satisfied:*

1 for each $i \in I$, the preferred set $P_i(\overline{x})$ is convex and $\overline{x}_i \in \operatorname{cl} P_i(\overline{x})$,

2 the total production set Y is convex.

Proof. Set $U = \sum_{i\in I} P_i(\overline{x}) - Y - \omega - Z$. Replacing ω by $\sum_{i\in I} \overline{x}_i - \sum_{j\in J} \overline{y}_j - z$, we get $U = \sum_{i\in I}(P_i(\overline{x}) - \overline{x}_i) - \sum_{j\in J}(Y_j - \overline{y}_j) + (z - Z)$. It follows from Conditions 1 and 2 and the convexity of Z that the set U is convex. Since the consumption allocation $\overline{x}$ is weakly Pareto optimal, we have $0 \notin U$. Using the finite dimensional separation theorem for disjoint convex sets, let $p \in \mathbb{R}^\ell$, $p \neq 0$ be such that $p \cdot U \geq 0$, that is,

$$p \cdot \Big(\sum_{i\in I}(P_i(\overline{x}) - \overline{x}_i) - \sum_{j\in J}(Y_j - \overline{y}_j) + (z - Z)\Big) \geq 0.$$

Using Condition (1), it is readily seen that for each $i \in I$,

$$x_i \in P_i(\overline{x}) \Longrightarrow p \cdot x_i \geq p \cdot \overline{x}_i.$$

We have also for each $j \in J$,

$$y_j \in Y_j \Longrightarrow p \cdot y_j \leq p \cdot \overline{y}_j$$

and $p \cdot (z - Z) \geq 0$. From the last relation and the facts that Z is a cone and $z \in Z$ it follows that

$$p \in Z^0 \text{ and } p \cdot z = p \cdot (\sum_{i\in I} \overline{x}_i - \omega - \sum_{j\in J} \overline{y}_j) = 0.$$

□

For a weakly Pareto optimal consumption allocation x and the corresponding production allocation y verifying $\sum_{i\in I} x_i - \sum_{j\in J} y_j \in \omega + Z$, a nonzero price p such that (x, y) is a quasiequilibrium relative to p is referred to as a **supporting price**. Proposition 4.3.3 can be shortly rephrased:

Under the conditions (1) and (2), a weakly Pareto optimal consumption allocation is price supported.

Obviously, the same holds true when a nonsatiation assumption replaces the local nonsatiation assumed in (1), if $\overline{x}$ is a weakly Pareto optimal allocation for augmented preferences (in particular if $\overline{x}$ is a weakly Pareto optimal allocation for strictly quasi-convex preferences).

COROLLARY 4.3.1 *Let $\overline{x}$ be a weakly Pareto consumption allocation of $\mathcal{E}$ for augmented preferences and let $\overline{y} \in \prod_{j\in J} Y_j$ be such that $\sum_{i\in I} \overline{x}_i - \omega - \sum_{j\in J} \overline{y}_j = z \in Z$. There is a nonzero price system p such that $(\overline{x}, \overline{y})$ is a quasiequilibrium relative to p, provided the following conditions are satisfied:*

1 *for each $i \in I$, the preferred set $P_i(\overline{x})$ is convex and $P_i(\overline{x}) \neq \emptyset$,*

2 *the total production set Y is convex.*

Proof. It suffices to apply Proposition 4.3.3 to the economy

$$\mathcal{E}' = \Big(\langle \mathbb{R}^\ell, \mathbb{R}^\ell \rangle, (X_i, \widehat{P}_i, \omega_i)_{i\in I}, (Y_j)_{j\in J}, (\theta_{i,j})_{\substack{i\in I \\ j\in J}}, Z\Big)$$

Let $p \neq 0$ which supports $(\overline{x})$. For each $i \in I$, from $p \cdot \widehat{P}_i(\overline{x}) \geq p \cdot \overline{x}_i$ we immediately deduce $p \cdot P_i(\overline{x}) \geq p \cdot \overline{x}_i$. □

4.3.2 Decentralizing fuzzy core allocations

Let us now start with $\overline{x} \in \mathcal{C}^f(\mathcal{E})$.

PROPOSITION 4.3.4 *Let $\overline{x}$ be an element of the fuzzy core of $\mathcal{E}$ and let $\overline{y} \in \prod_{j\in J} Y_j$ be such that $\sum_{i\in I} \overline{x}_i - \omega - \sum_{j\in J} \overline{y}_j = z \in Z$. There is a nonzero price system p such that $(\overline{x}, \overline{y}, p)$ is a quasiequilibrium of $\mathcal{E}$, provided the following conditions are satisfied:*

1 *for each $i \in I$, the preferred set $P_i(\overline{x})$ is convex and $\overline{x}_i \in \operatorname{cl} P_i(\overline{x})$,*

2 *for each $j \in J$, the production set Y_j is convex.*

Proof. Define $G = \operatorname{co}\Big(\bigcup_{i\in I} (P_i(\overline{x}) - \sum_{j\in J} \theta_{i,j} Y_j - \omega_i)\Big)$. We first claim that $G \cap Z = \emptyset$. Indeed, by Conditions (1) and (2), each

$$P_i(\overline{x}) - \sum_{j\in J} \theta_{i,j} Y_j - \omega_i$$

is nonempty and convex. So, by an elementary convexity result,

$$G = \Big\{\sum_{i\in I}\lambda_i(x_i - \sum_{j\in J}\theta_{i,j}y_j - \omega_i) : x_i \in P_i(\overline{x}),\ y_j \in Y_j,\ \lambda_i \geq 0,\ \sum_{i\in I}\lambda_i = 1\Big\}.$$

If $G \cap Z \neq \emptyset$, there exists $\lambda = (\lambda_i)_{i\in I}$ with $\lambda_i \geq 0$, $\sum_{i\in I}\lambda_i = 1$, and $x \in \prod_{\{i\in I:\lambda_i>0\}} X_i = X_\lambda$ such that

$$\sum_{\{i\in I:\lambda_i>0\}}\lambda_i(x_i - \omega_i) \in \sum_{i\in I}\lambda_i\sum_{j\in J}\theta_{i,j}Y_j + Z, \text{ and} \tag{4.13}$$

$$x_i \in P_i(x) \text{ for each } i \text{ such that } \lambda_i > 0. \tag{4.14}$$

One recognizes in (4.13) and in (4.14) the relations (4.5) and (4.6) which insures in Definition 4.1.5 that the fuzzy coalition $\lambda = (\lambda_i)_{i\in I}$ blocks with $x \in X_\lambda$ the attainable consumption allocation $\overline{x}$. This contradicts the fact that $\overline{x} \in \mathcal{C}^f(\mathcal{E})$. We thus have proved that $G \cap Z = \emptyset$.

Applying the finite dimensional separation theorem for disjoint convex sets, let $p \neq 0$ in $\mathbb{R}^\ell$ and $\alpha \in \mathbb{R}$ be such that $p \cdot Z \leq \alpha \leq p \cdot G$. Since Z is a cone, $\alpha \geq 0$ and $p \cdot Z \leq 0 \leq p \cdot G$. From this, we deduce:

$$x_i \in P_i(\overline{x}) \Longrightarrow p \cdot x_i \geq p \cdot \omega_i + \sum_{j\in J}\theta_{i,j}p \cdot Y_j. \tag{4.15}$$

Using Condition (1), we have for each $i \in I$:

$$p \cdot \overline{x}_i \geq p \cdot \omega_i + \sum_{j\in J}\theta_{i,j}p \cdot Y_j, \text{ and, in particular,} \tag{4.16}$$

$$p \cdot \overline{x}_i \geq p \cdot \omega_i + \sum_{j\in J}\theta_{i,j}p \cdot \overline{y}_j. \tag{4.17}$$

Summing over $i \in I$, we get $p \cdot \sum_{i\in I}(\overline{x}_i - \omega_i) \geq p \cdot \sum_{j\in J}\overline{y}_j$. From $\sum_{i\in I}(\overline{x}_i - \omega_i) - \sum_{j\in J}\overline{y}_j \in Z$, we deduce $p \cdot \sum_{i\in I}(\overline{x}_i - \omega_i) = p \cdot \sum_{j\in J}\overline{y}_j$, and, finally, for each $i \in I$,

$$p \cdot \overline{x}_i = p \cdot \omega_i + \sum_{j\in J}\theta_{i,j}p \cdot \overline{y}_j. \tag{4.18}$$

Using (4.15) and (4.18) we get:

$$x_i \in P_i(\overline{x}) \Longrightarrow p \cdot x_i \geq p \cdot \overline{x}_i = p \cdot \omega_i + \sum_{j\in J}\theta_{i,j}p \cdot \overline{y}_j.$$

Let now $j \in J$ and $y_j \in Y_j$. Using (4.16) and (4.18) for i such that $\theta_{i,j} > 0$, we get $p \cdot y_j \leq p \cdot \overline{y}_j$. We have thus proved

$$y_j \in Y_j \Longrightarrow p \cdot y_j \leq p \cdot \overline{y}_j.$$

The two last implications, with $p \in Z^0$ and

$$p \cdot \Big(\sum_{i \in I} (\overline{x}_i - \omega_i) - \sum_{j \in J} \overline{y}_j \Big) = 0,$$

complete the proof. □

An important byproduct of this result is that, combined with the assumptions of the nonemptiness theorem stated in Proposition 4.2.6, it gives a complete set of sufficient assumptions for the existence of quasiequilibrium for the private ownership economy $\mathcal{E}$.

PROPOSITION 4.3.5 *Under the following assumptions:*

1 The set $\widehat{X}$ is compact. Moreover, for each $i \in I$,

- *(a) X_i is convex,*
- *(b) For each $x_i \in X_i$, $(P_i)^{-1}(x_i)$ is open in $\prod_{k \in I} X_i$,*
- *(c) For each $x \in \widehat{X}$, $P_i(x)$ is convex with $x_i \notin P_i(x)$, $x_i \in \operatorname{cl} P_i(x)$,*
- *(d) $\omega_i \in X_i - \sum_{j \in J} \theta_{i,j} Y_j - Z$;*

2 For each $j \in J$, Y_j is convex;

3 Z is a convex cone contained in the negative orthant $\mathbb{R}^{\ell}_{-}$;

4 For each i, for every $x \in \widehat{X}$, if $z_i \in P_i(x)$ and $v_i \in X_i$ then there exists $\lambda : 0 \leq \lambda < 1$ such that $(\lambda z_i + (1-\lambda) v_i) \in P_i(x)$.

the economy $\mathcal{E}$ under consideration in this chapter has a quasiequilibrium.

This quasiequilibrium is nontrivial if $\omega \in \operatorname{int}(\sum_{i \in I} X_i - Y - Z)$.

Proof. By Assumptions 1, 2, 3, 4 and Proposition 4.2.6, there exists an attainable consumption allocation $\overline{x} \in \mathcal{C}^f(\mathcal{E})$. By Assumptions 1(c) and 2, and the previous proposition, there exist $\overline{y} \in \prod_{j \in J} Y_j$ and $\overline{p} \neq 0$ such that $(\overline{x}, \overline{y}, \overline{p})$ is a quasiequilibrium of $\mathcal{E}$. The proof of the last assertion is obvious. □

4.3.3 From quasiequilibrium to equilibrium

Let us start with the quasiequilibrium $(\overline{x}, \overline{y}, \overline{p})$ obtained in Proposition 4.3.5. If it is nontrivial, this quasiequilibrium is an equilibrium under the different sets of assumptions studied in the last section of Chapter 2. To see that, recall the definition for each i of augmented preferences:

$$\widehat{P}_i(x) = \{x'_i \in X_i : x'_i = x_i + \lambda(x''_i - x_i), 0 < \lambda \leq 1, \ x''_i \in P_i(x)\}.$$

For each $i \in I$, if $x_i \in \widehat{P}_i(\overline{x})$ then $x_i = \overline{x}_i + \lambda(x'_i - \overline{x}_i)$ for some $\lambda \geq 1$ and $x'_i \in P_i(\overline{x})$. It follows that $\overline{p} \cdot x_i = \overline{p} \cdot \overline{x}_i + \lambda(\overline{p} \cdot x'_i - \overline{p} \cdot \overline{x}_i) \geq \overline{p} \cdot \overline{x}_i$. In other words, $(\overline{x}, \overline{y}, \overline{p})$ is a quasiequilibrium of the economy $\mathcal{E}'$ deduced from $\mathcal{E}$ by replacing preferences P_i by augmented preferences $\widehat{P}_i$. Since $(\overline{x}, \overline{y}, \overline{p})$ is a nontrivial quasiequilibrium of $\mathcal{E}$, it is also a nontrivial quasiequilibrium of $\mathcal{E}'$. We have already noticed in Chapter 2 (see Remark 2.3.1) that if Assumption 4 is satisfied for correspondences P_i then it is also satisfied for correspondences $\widehat{P}_i$. It follows that, in view of Assumption 4, any irreducibility assumption stated in Chapter 2 implies that $(\overline{x}, \overline{y}, \overline{p})$ is an equilibrium of $\mathcal{E}'$, thus an equilibrium of $\mathcal{E}$.

From the point of view of the generality of the model and the continuity assumptions made on preferences, the quasiequilibrium existence result obtained in this chapter is intermediate between the ones obtained in Chapter 2 and 3. Its other assumptions have to be compared with the ones used in Proposition 2.2.2 in Chapter 2 and the ones used in Proposition 3.2.3 in Chapter 3. There is no closedness assumption on the individual consumption sets, the total production sets and on $Y + Z$ in the previous proposition. However, these assumptions are implicit in the assumption that $\widehat{X}$ is compact. Assumption 4, used here for establishing the existence of a quasiequilibrium, was only used in Chapters 2 and 3 for passing from quasiequilibrium to equilibrium. The most noticeable difference is in the local nonsatiation at every component of an attainable consumption allocation assumed in the previous proposition while it was proved in Chapters 2 and 3 that nonsatiation is sufficient for the quasiequilibrium existence of an economy defined on a finite dimensional commodity space.

Chapter 5

INFINITE DIMENSIONAL ECONOMIES

Recall that in general equilibrium a commodity is defined not only by its physical properties but also by the date, the location and the state of nature that precise the conditions of its availability. The classical general equilibrium model hypothesis of a finite number of commodities, maintained in Chapters 2, 3 and 4, implies that the economic activity extends over only finitely many dates, locations and events. Such an hypothesis, contrary to the framework of many applied models of location, growth and finance, limits dramatically the application of the results to understanding the real economic life.

Moreover, the physical properties of a commodity can vary continuously depending on some characteristics. In such a setting, a commodity is defined as a point in a space of characteristics describing, for instance, its location or its content of more fundamental elements. The space of characteristics is equipped with a metric defining closeness of commodities. The definition of commodity bundles as measures on this space corresponds to the idea that commodities with close characteristics should be considered as good substitutes.

To give solid foundations to an analysis of allocation over time or states of nature was the first motivation for the study of infinite dimensional economies ([148], [62], [34], [35]). Financial economics first, then the analysis of commodity differentiation ([128], [112], [109]) added their own difficulties. Infinite dimensional economies are nowadays required to allow to formalize very diverse settings, ideally any conceivable economic setting.

The ambition of this chapter is necessarily more limited. Keeping in mind the three previous specific examples of infinite dimensional economies, we intend to give a solution to the equilibrium existence problem general enough to have implications for the equilibrium existence problem in these three frameworks and the more general settings related to them.

In a preliminary section, we motivate the mathematical structures used in this chapter, starting from a list of the main commodity and price spaces used in economic analysis. We then state the equilibrium and optimality concepts appropriate to infinite dimensional analysis.

We next face the equilibrium existence problem in infinite dimensional economies. As we will see, the infinite dimensionality of the commodity space does not prevent to get existence of Edgeworth equilibria under conditions similar to the ones used in the finite dimensional case. The same is not true for the existence of decentralizing prices. Since such a decentralization is based on a separation argument applied to two appropriately defined disjoint convex sets, the existence of decentralizing prices meets here the limits put to the existence of a closed separating hyperplane by the Hahn–Banach theorem which requires in an infinite dimensional setting an interiority nonemptiness condition for one of both sets. Because the approach used in Chapter 4 for finite dimensional economies allows for separating these two different questions in the equilibrium existence problem, we have chosen to use here the same approach and to give first in Section 5.2 sufficient conditions for the existence of Edgeworth equilibria before studying in Section 5.3 the difficult problem of their decentralization.

As we shall see, this decentralization can be based either on an interiority assumption on the characteristics of the economy or on the so-called properness assumptions made on preferred and production sets. The interest of properness assumptions is to open a room for equilibrium existence theorems in ordered topological vector spaces whose positive cone has an empty interior. A more and more acute statement of these assumptions and of their conditions of application is the decisive progress of the last twenty years, initiated by the paper of Mas-Colell [130] published in 1986, first circulated in 1983.

We will use in this chapter two different versions of properness. The first one, uniform properness, yields equilibrium existence theorems for exchange and production economies defined on topological vector lattices as commodity spaces. As an additional restriction, consumption sets of the agents are assumed to coincide with the positive cone of the commodity space. Further in the chapter, we show how the requirement of uniform properness can be reduced to pointwise properness for an exchange economy (with consumption sets equal to the positive cone) defined on a lattice commodity space whose positive cone has a strictly positive element.

A second version of properness allows for equilibrium existence theorems under weaker assumptions on the commodity space, more convenient for studying commodity differentiation and in some other settings. With this version of properness, it is also possible to drop the unpleasant assumption that consumption sets coincide with the positive cone.

For the other assumptions on the primitives of the economy, we work in the whole chapter at the same level of generality as in Chapter 4 for the finite dimensional case. In particular, preferences are nonordered, they may depend on the consumption of the other consumers, and we do not assume any kind of monotonicity. Production sets need not satisfy free-disposal.

A last section is devoted to a brief historical survey.

5.1 Preliminaries

5.1.1 Commodities and prices

Assume that from the point of view of their physical properties there are ℓ different goods (or services). A commodity bundle should specify the quantity of each good depending on time, location or state of nature. Such a naive statement suggests that a commodity bundle should be a vector valued function and that an admissible commodity space should be a function space. Actually, on a measure space (Ω, Σ, μ) formalizing time or uncertainty or on the interval $[0,1]$ with its Lebesgue measure formalizing continuous time, spaces $L_p^\ell(\Omega, \Sigma, \mu)$, $L_p^\ell([0,1])$, $1 \leq p \leq \infty$ are the most currently used commodity spaces in economics. As usual, let us consider as a same function two measurable vector valued functions from Ω to $\mathbb{R}^\ell$ which coincide μ-almost everywhere in Ω (resp. two measurable vector valued functions from $[0,1]$ to $\mathbb{R}^\ell$ which coincide almost everywhere). Spaces $L_\infty^\ell(\Omega, \Sigma, \mu)$ (resp. $L_\infty^\ell[0,1]$) that are spaces of (classes of) essentially bounded measurable functions from Ω to $\mathbb{R}^\ell$ (resp. from $[0,1]$ to $\mathbb{R}^\ell$) are relevant to the allocation of resources over time or states of nature. Spaces $L_p^\ell(\Omega, \Sigma, \mu)$ (resp. $L_p^\ell[0,1]$), $p \geq 1$, are spaces of (classes of equivalence of) of p-integrable measurable functions from Ω to $\mathbb{R}^\ell$ (resp. from $[0,1]$ to $\mathbb{R}^\ell$). Spaces $L_p^\ell(\Omega, \Sigma, \mu)$ arise in Finance and in any uncertainty setting where mean, variance and eventually the moments of superior order of commodity bundles matter. Spaces $L_p^\ell(\Omega, \Sigma, \mu)$ (resp. $L_p^\ell[0,1]$) arise in any intertemporal setting where p-integrability of commodity bundles matters.[1]

More generally, in order to model intertemporal allocation problems under uncertainty, one can equip the σ-algebra Σ of the measure space (Ω, Σ, μ) with an increasing time filtration $\{\Sigma_t\}_{t\geq 0}$, where Σ_t denotes the set of events corresponding to the information available at time t. Commodity bundles are stochastic processes adapted to the filtration and the commodity space is some L_p-type space of processes on $(\Omega, \Sigma, \{\Sigma_t\}_{t\geq 0}, \mu)$. In discrete ($t \in \mathbb{N}$) or continuous time ($t \in [0,T]$), such a framework is especially used in Finance (For more details, see Duffie [66] and its references).

A noticeable exception to the previous modellings of commodity bundles is given by the space $M(K)$ of all finite countably additive signed measures on a compact metric space of characteristics used in models of location or for the analysis of commodity differentiation. In this setting, a point of K is to be understood as a complete description of a unit of a commodity. A commodity bundle is not a function from K to some vector space but a finite measure on K, that is a countably additive function $m\colon \mathcal{B}(K) \to \mathbb{R}$ from the Borel σ-field of subsets of K, $\mathcal{B}(K)$, to $\mathbb{R}$; for $B \in \mathcal{B}(K)$, $m(B)$ denotes the total number of units of commodities having characteristics in B. In this formalization, Dirac measures δ_t, defined for each $t \in K$ by $\delta_t(B) = 1$ if $t \in B$, 0 otherwise, play a special role and have a special interpretation.

Notice immediately that, when endowed with their canonical ordering, the previous examples of commodity spaces are ordered vector spaces and more precisely vector lattices or Riesz spaces. Indeed, two measurable vector valued functions x and y from (Ω, Σ, μ) to $\mathbb{R}^\ell$ are said to satisfy $x \leq y$ if and only if for the canonical order on $\mathbb{R}^\ell$ one has $x(\omega) \leq y(\omega)$ μ-almost everywhere in Ω. This relation is compatible with the vector operations. It is an order relation if two functions which coincide μ-almost everywhere in Ω are considered as a same function. With this convention, all spaces $L_p(\Omega, \Sigma, \mu)$, $1 \leq p \leq \infty$, are ordered by the previous relation. Finally, all spaces L_p inherit the lattice structure of $\mathbb{R}^\ell$. For a couple (x, y) of measurable vector valued functions x and y from (Ω, Σ, μ) to $\mathbb{R}^\ell$, one define the supremum $x \vee y$ by $(x \vee y)(\omega) = x(\omega) \vee y(\omega)$

[1] See for instance the commodity space used for modeling the storage-release production technology in [65, Example 6].

μ-almost everywhere in Ω and it can be shown that if x and y belong to some L_p then the same is true for $x \vee y$.

On the other hand, as functions from the Borel subsets of K to $\mathbb{R}$, the elements of $M(K)$ are ordered using the natural order of $\mathbb{R}$ by $x \leq y$ if and only if $x(A) \leq y(A)$ for every Borel subset of K. As well-known, $M(K)$ is lattice-ordered. A classical reference for the properties of spaces L_p, $M(K)$ and more generally many other special spaces is Dunford and Schwartz [68].

In this chapter, we will stay at an abstract theoretical level sufficiently general to encompass as particular instances the previous commodity spaces. The commodity space L will always be a vector space that we will equip with the additional structures corresponding to the needs of the mathematical tools used in the proof of each desired result. As, in particular, we want to formalize the presence of disposal economic activities, we shall assume that the commodity space is an ordered vector space. This assumption would not be necessary for defining and in certain cases proving the existence of a no-disposal (or exact) equilibrium. But let us recall that free-disposal assumptions have a long-standing history in economics. Moreover, as in the finite dimensional case, the introduction of disposal activities will allow us to state in a same result existence of no-disposal and free-disposal equilibrium.

A price vector, a "valuation" in the terminology of Debreu [56], is a linear functional on the commodity space, that is, an element of some vector space L' that we require in this chapter to be in duality with the commodity space L. By this requirement, we mean that there exists a bilinear functional $(x, p) \mapsto \langle x, p \rangle$ from $L \times L'$ to $\mathbb{R}$ such that for any $x \neq 0$ in L there is some $p \in L'$ such that $\langle x, p \rangle \neq 0$ and for any $p \neq 0$ in L' there is some $x \in L$ such that $\langle x, p \rangle \neq 0$. This duality makes of L' a vector subspace of the algebraic dual L^* of L satisfying that for each nonzero $x \in L$ there is some $p \in L'$ such that $p(x) \neq 0$.

When the commodity space is the finite dimensional commodity space $\mathbb{R}^\ell$, the algebraic dual of $\mathbb{R}^\ell$, identified with the same space $\mathbb{R}^\ell$, has no proper subspace in duality with $\mathbb{R}^\ell$; it is also the space of all linear functionals on $\mathbb{R}^\ell$ continuous for its unique Hausdorff vector space topology. For being a price space when $\mathbb{R}^\ell$ is the commodity space, the economically meaningful space $\mathbb{R}^\ell$ is thus the unique candidate. Moreover, an equilibrium price is automatically continuous.

When the commodity space is an infinite dimensional vector space L, several vector subspaces of the algebraic dual L^* of L can be put in duality with L. The choice of a price space L' determines the properties which will be satisfied by a price system as well as the compatible topologies[2] on the commodity space L which will be used for formulating assumptions on the economy.

In order to have an economic interpretation in term of prices of individual commodities, a price system should belong to a function space when the commodity space is a function space, to a space of continuous real functions on K if the commodity space is $M(K)$. The last continuity requirement for prices is a consequence of the assumed substituability of commodities with nearly the same description in the space K. This suggests that L_q-type spaces (where q is such that $\frac{1}{q}+\frac{1}{p}=1$) are admissible price spaces for L_p-type commodity spaces, when $1 \leq p < \infty$. The topological duals of L_∞ and $M(K)$ equipped with their norm topology are of more difficult economic interpretation, so that one should prefer their pre-duals[3] L_1 and $C(K)$, the space of all continuous real functions defined on K. This choice will cause specific difficulties in the equilibrium existence proofs.

Following a terminology introduced by Aliprantis, Brown and Burkinshaw [7], the couple $\langle L, L'\rangle$ will be called **commodity–price duality**. Given a price system $p \in L'$, the value $\langle x, p\rangle$ of a commodity bundle $x \in L$ will be denoted by $p \cdot x$. We will state in the sequel different assumptions on the commodity–price duality. In order to keep the exposition at the level of an initiation to infinite dimensional economies, we shall stay sometimes below the full generality of the mathematical structures that could be used for obtaining equilibrium existence results.

5.1.2 Equilibrium and optimality concepts

With $\langle L, L'\rangle$ as commodity–price duality, a private ownership economy can be specified by

$$\mathcal{E} = \Big(\langle L, L'\rangle, (X_i, P_i, \omega_i)_{i\in I}, (Y_j)_{j\in J}, (\theta_{i,j})_{\substack{i\in I\\ j\in J}}, Z\Big)$$

where I is a finite set of consumers, J is a finite set of producers. For each consumer $i \in I$, $X_i \subset L$ and $\omega_i \in L$ are respectively the *consumption set* and the *initial endowment* of consumer i. As in Chapter 4, the *preference correspondence* $P_i\colon \prod_{k\in I} X_k \to X_i$ defines preferred sets for i which depend on the current consumption of other consumers but not

[2]A topology τ on L is compatible with the duality $\langle L, L'\rangle$ if $L' = (L, \tau)'$ (the topological dual of L), that is, if L' is the vector space of all τ-continuous linear functionals on L.

[3]L_∞ is the norm-dual of L_1 when the underlying measure space is σ-finite.

on production nor on prices. For each producer $j \in J$, $Y_j \subset L$ is the *production set* of j, and $\theta_{i,j}$ is the *contractual claim* of consumer i on the profit of j. Let L_+ denote the positive cone of L. The convex *disposal cone* Z is contained in $L_- := -(L+)$.

With these precisions, the notation adopted for the valuation of commodity bundles and an obvious adaptation of the definition of the polar cone

$$Z^0 := \{p \in L' : p \cdot z \leq 0, \ \forall z \in Z\}$$

of the disposal cone Z, quasiequilibrium and equilibrium definitions can mimic the definitions given in the finite dimensional case.

DEFINITION 5.1.1 *An* ***allocation*** $(x, y) \in \prod_{i \in I} X_i \times \prod_{j \in J} Y_j$ *is called* ***feasible*** (*or* ***attainable***) *if* $\sum_{i \in I} x_i - \sum_{j \in J} y_j - \sum_{i \in I} \omega_i \in Z$.

DEFINITION 5.1.2 *A t-uple* $(\overline{x}, \overline{y}, \overline{p})$ *consisting of an attainable allocation* $(\overline{x}, \overline{y})$ *and a nonzero price vector* $\overline{p} \in L'$ *is said to be an* ***equilibrium*** (*resp. a* ***quasiequilibrium***) *if*

1 For each $i \in I$, $\overline{p} \cdot \overline{x}_i \leq \overline{p} \cdot \omega_i + \sum_{j \in J} \theta_{i,j} \overline{p} \cdot \overline{y}_j$ *and*

$$x_i \in P_i(\overline{x}) \Longrightarrow \overline{p} \cdot x_i > \overline{p} \cdot \overline{x}_i \ (\textit{resp. } x_i \in P_i(\overline{x}) \Longrightarrow \overline{p} \cdot x_i \geq \overline{p} \cdot \overline{x}_i),$$

2 For each $j \in J$, *for all* $y_j \in Y_j$, $\overline{p} \cdot y_j \leq \overline{p} \cdot \overline{y}_j$,

3 $\overline{p} \in Z^0$, *and* $\overline{p} \cdot \sum_{i \in I} \overline{x}_i = \overline{p} \cdot \sum_{j \in J} \overline{y}_j + \overline{p} \cdot \sum_{i \in I} \omega_i$.

Consequently, for every $i \in I$, $\overline{p} \cdot \overline{x}_i = \overline{p} \cdot \omega_i + \sum_{j \in J} \theta_{i,j} \overline{p} \cdot \overline{y}_j$.

DEFINITION 5.1.3 *A quasiequilibrium* $(\overline{x}, \overline{y}, \overline{p})$ *is said to be* ***nontrivial*** *if for some* $i \in I$ *and for some* $x_i \in X_i$, $\overline{p} \cdot x_i < \overline{p} \cdot \overline{x}_i$.

Maybe more than in the finite dimensional case, it should be emphasized that equilibrium (resp. quasiequilibrium) is the result of decentralized decisions made at a same date which is the "ideal" origin of time, before any resolution of uncertainty, by individual agents who have a full information on all possible present and future events.

As far one deals only with optimality concepts, it is not necessary to precise the price space. The optimality concepts of Chapter 4 do not need any adaptation. We summarize them in the following definitions where we use the same notations. In particular,

- $\omega = \sum_{i \in I} \omega_i$ is the total initial endowment;
- $Y = \sum_{j \in J} Y_j$ is the total production set;
- $\mathcal{A}(\mathcal{E}) = \left\{(x, y) \in \prod_{i \in I} X_i \times \prod_{j \in J} Y_j : \sum_{i \in I} x_i - \sum_{j \in J} y_j - \omega \in Z\right\}$ is the set of all attainable allocations;
- $\widehat{X} = \left\{x \in \prod_{i \in I} X_i : \exists y \in Y, \sum_{i \in I} x_i - y - \omega \in Z\right\}$ denotes the set of all attainable consumption allocations.

DEFINITION 5.1.4 *An attainable consumption allocation $x \in \widehat{X}$ is*

1 ***weakly Pareto optimal*** *if there is no $x' \in \widehat{X}$ satisfying $x'_i \in P_i(x)$ for each $i \in I$;*

2 *a* ***core allocation*** *if it is not blocked by any coalition in the sense that there is no $S \subset I$, $S \neq \emptyset$, and $x_S \in \prod_{i \in S} X_i$ such that*

   (a) $\sum_{i \in S}(x_{i,S} - \omega_i) \in \sum_{i \in S} \sum_{j \in J} \theta_{i,j} Y_j + Z$,

   (b) $x_{i,S} \in P_i(x)$ *for each* $i \in S$;

3 *an* ***Edgeworth allocation*** *if for any integer r, the r-replica of x belongs to the core $\mathcal{C}^r(\mathcal{E})$ of the r-replica of $\mathcal{E}$.*

Let $\mathcal{T} = [0,1]^I \setminus \{0\}$ be the set of all **fuzzy coalitions**, that is, the set of all nonzero vector of rates of participation. For each $t \in \mathcal{T}$, let $X_t = \prod_{\{i:\, t_i > 0\}} X_i$, and

$$\widehat{X}_t := \left\{x_t = (x_{i,t}) \in X_t \colon \sum_{i \in I} t_i x_{i,t} - \sum_{i \in I} t_i \omega_i \in \sum_{i \in I} t_i \sum_{j \in J} \theta_{i,j} Y_j + Z\right\}.$$

If $x \in \widehat{X}$,

$$P_t(x) = \left\{x_t = (x_{i,t}) \in X_t \colon x_{i,t} \in P_i(x) \text{ for each } i \text{ such that } t_i > 0\right\}.$$

The set $\widehat{X}_t$ can be interpreted as the set of all attainable consumption assignments for the fuzzy coalition t. The sets $P_t(x)$ define a correspondence $P_t \colon \prod_{i \in I} X_i \to X_t$ interpreted as a preference correspondence for the fuzzy coalition t.

For any positive integer $r \geq 1$, let us consider the set

$$\mathcal{T}_r := \{t = (t_i)_{i \in I}\colon\ rt_i \in \{0, 1, \dots, r\} \text{ for each } i \in I\}.$$

Exactly as in Chapter 4, it is easily seen that if each X_i, Y_j and every $P_i(x)$ are convex, then an attainable consumption allocation $x \in \widehat{X}$ belongs to $\mathcal{C}^r(\mathcal{E})$ if there is no $t \in \mathcal{T}_r$ and $x_t \in \widehat{X}_t \cap P_t(x)$. Consequently, under the same convexity assumptions, an attainable consumption allocation $x \in \widehat{X}$ is an Edgeworth equilibrium if and only if there is no fuzzy coalition $t \in \mathcal{T} \cap \mathbb{Q}^I$ with rational rates of participation and $x_t \in \widehat{X}_t \cap P_t(x)$.

The previous considerations motivate the following definition.

DEFINITION 5.1.5 *A fuzzy coalition $t \in \mathcal{T}$* ***blocks*** *an attainable consumption allocation $x \in \widehat{X}$ if there is some $x_t \in \widehat{X}_t \cap P_t(x)$.*

The ***fuzzy core*** *of the economy $\mathcal{E}$ is the set $\mathcal{C}^f(\mathcal{E})$ of all attainable consumption allocations which are blocked by no fuzzy coalition.*

As just observed, under the convexity assumptions precised above and which will be made in the next subsection, the set $\mathcal{C}^r(\mathcal{E})$ is the set of all attainable consumption allocations $x \in \widehat{X}$ which are blocked by no fuzzy coalition $t \in \mathcal{T}_r$.

5.2 Edgeworth equilibrium existence and nonemptiness of the fuzzy core

In this section, **the ordered commodity vector space is assumed to be equipped with a Hausdorff vector space topology** σ. We will prove the nonemptiness results under assumptions on $\mathcal{E}$ which parallel the assumptions used in Chapter 4. The notations are the same. The product set $X = \prod_{k \in I} X_k$ is endowed with the topology induced by the product topology on L^I that we note σ^I.

1 The set $\widehat{X}$ is σ^I-compact. Moreover, for each $i \in I$,

(a) X_i is convex,

(b) For each $x_i \in X_i$, $(P_i)^{-1}(x_i)$ is σ^I-open in X,

(c) For each $x \in \widehat{X}$, $P_i(x)$ is convex with $x_i \notin P_i(x)$,

(d) $\omega_i \in X_i - \sum_{j \in J} \theta_{i,j} Y_j - Z$;

2 For each $j \in J$, Y_j is convex;

3 Z is a convex cone contained in the negative cone $L_- = -(L_+)$.

By considering traces of economy $\mathcal{E}$ on finite dimensional subspaces of the commodity space, we will first prove an infinite dimensional analogue of Corollary 4.2.1 to Proposition 4.2.4 in Chapter 4.

PROPOSITION 5.2.1 *Let r be any positive integer. Under the assumptions 1, 2, 3 stated above on the economy $\mathcal{E}$, the set $\mathcal{C}^r(\mathcal{E})$ is nonempty.*

Proof. By Assumption 1(d), for each $i \in I$ and $j \in J$ there exist $\underline{x}_i \in X_i$, $\underline{y}_{i,j} \in Y_j$, $\underline{z}_i \in Z$ such that $\underline{x}_i = \omega_i + \sum_{j\in J} \theta_{i,j}\underline{y}_{i,j} + \underline{z}_i$. Let $\mathcal{F}$ be the collection of all finite dimensional subspaces of L containing $\underline{x}_i$, $\underline{y}_{i,j}$, $\underline{z}_i$ for all $i \in I$, $j \in J$.

For each $F \in \mathcal{F}$, we set $X_{i,F} = X_i \cap F$; $X_F = \prod_{i\in I} X_{i,F}$; if $x \in X_F$, $P_{i,F}(x) = P_i(x) \cap X_{i,F}$; $Y_{j,F} = Y_j \cap F$; $Y'_F = (Y+Z) \cap F$; $Z_F = Z \cap F$. We observe that Z_F can be considered as a disposal cone for the economy

$$\mathcal{E}_F = ((X_{i,F}, P_{i,F}, \omega_i)_{i\in I}, (Y_{j,F})_{j\in J}, (\theta_{i,j})_{\substack{i\in I\\ j\in J}}, Z_F)$$

defined on the commodity space F if the order on F is the order induced by the order on L. On the other hand, it is easily seen that Y'_F contains $\sum_{j\in J} Y_{j,F} + Z_F$ and that

$$X'_F := \{x \in \prod_{i\in I} X_{i,F}\colon \sum_{i\in I} x_i \in \sum_{i\in I} \omega_i + Y'_F\} = \widehat{X} \cap F^I.$$

Thus X'_F is compact if F is equipped with the topology induced by the topology σ of L and F^I with the product of these topologies,[4] and the finite dimensional economy $\mathcal{E}_F$ satisfies the assumptions of Proposition 4.2.4 in Chapter 4. In view of its conclusion, let $\overline{x}_F \in \widehat{X} \cap F^I$ be such that $\widehat{X}_{t,F} \cap P_{t,F}(\overline{x}_F) = \emptyset$ for every $t \in \mathcal{T}_r$.

Notice that the collection $\mathcal{F}$, ordered by inclusion, is directed.[5] Since $\widehat{X}$ is σ^I-compact, by passing to subnets if necessary, we can assume that $\overline{x}_F \xrightarrow{\sigma} \overline{x} \in \widehat{X}$. To end the proof, we show by contraposition that $\overline{x} \in \mathcal{C}^r(\mathcal{E})$. Indeed, if for some $t \in \mathcal{T}_r$ there exists $x_t \in \widehat{X}_t \cap P_t(\overline{x})$ then, using Assumption 1(b) and the fact that the collection $\mathcal{F}$ is directed, we can find $F_0 \in \mathcal{F}$ such that

$$F \in \mathcal{F} \text{ and } F_0 \subset F \Longrightarrow x_t \in \widehat{X}_{t,F} \cap P_{t,F}(\overline{x}_F).$$

[4] Recall that the topology induced on F by the vector space topology σ is the unique Hausdorff vector space topology on F.

[5] Namely, for every couple (F_1, F_2) of elements of $\mathcal{F}$ there is some $F_0 \in \mathcal{F}$ such that $F_1 \subset F_0$ and $F_2 \subset F_0$.

This contradicts the assumption that $\widehat{X}_{t,F} \cap P_{t,F}(\overline{x}_F) = \emptyset$. □

The proofs of the next two results parallel the proofs given in Chapter 4 for a finite dimensional economy (See Proposition 4.2.5 and 4.2.6).

PROPOSITION 5.2.2 *Under the assumptions 1, 2, 3 stated above, the economy $\mathcal{E}$ has Edgeworth equilibria, that is, the set $\mathcal{C}^e(\mathcal{E})$ is nonempty.*

Proof. Recall that $\mathcal{C}^e(\mathcal{E}) = \bigcap_{r \geq 1} \mathcal{C}^r(\mathcal{E})$. It is easily seen that $\mathcal{C}^{r+1}(\mathcal{E}) \subset \mathcal{C}^r(\mathcal{E})$ for every $r \geq 1$. It follows from the definition and Assumption 1.(b) that each $\mathcal{C}^r(\mathcal{E})$ is a σ^I-closed subset of $\widehat{X}$, nonempty by the previous proposition. Then the nonemptiness of $\mathcal{C}^e(\mathcal{E})$ follows from the σ^I-compactness of $\widehat{X}$. □

In order to prove the nonemptiness of the fuzzy core, we posit now the same additional assumption as in Chapter 4.

4 For each i, for every $x \in \widehat{X}$, if $z_i \in P_i(x)$ and $v_i \in X_i$ then there exists $\lambda : 0 \leq \lambda < 1$ such that $(\lambda z_i + (1 - \lambda)v_i) \in P_i(x)$.

Let τ be a vector space topology on L, not necessarily related with the topology σ initially given on L. Assumption 4 is satisfied if for each i and for every $x \in \widehat{X}$ the consumption set X_i is convex and the preferred set $P_i(x)$ is τ-open in X_i. Assumption 4 is also satisfied if for each i, the consumption set X_i is convex and P_i is defined by a concave utility function.

We will use this additional assumption and the density of $\mathbb{Q}$, the set of rational numbers, in the set $\mathbb{R}$ of real numbers to prove the last nonemptiness result of this section.

PROPOSITION 5.2.3 *Under the assumptions 1, 2, 3, 4 stated above, the economy $\mathcal{E}$ has a nonempty fuzzy core, that is, $\mathcal{C}^f(\mathcal{E}) \neq \emptyset$.*

Proof. By the previous proposition, $\mathcal{C}^e(\mathcal{E}) \neq \emptyset$. Under the additional assumption 4, we now show that $\mathcal{C}^e(\mathcal{E}) \subset \mathcal{C}^f(\mathcal{E})$.

Let $x \in \mathcal{C}^e(\mathcal{E})$. If $x \notin \mathcal{C}^f(\mathcal{E})$, then there exist $t \in \mathcal{T}$ and $x_t \in X_t$ such that

$$\sum_{\{i \in I:\ t_i > 0\}} t_i x_{i,t} - \sum_{i \in I} t_i \omega_i \in \sum_{i \in I} t_i \sum_{j \in J} \theta_{i,j} Y_j + Z \tag{5.1}$$

$$x_{i,t} \in P_i(x) \quad \text{for each } i \text{ such that } t_i > 0. \tag{5.2}$$

Using (5.2), Assumption 4 and the convexity of $P_i(x)$, we can find some $\varepsilon > 0$ such that $0 \leq 1 - \varepsilon < \lambda \leq 1$ implies

$$\lambda x_{i,t} + (1-\lambda)\underline{x}_i \in P_i(x) \text{ for each } i \text{ such that } t_i > 0,$$

where as previously $\underline{x}_i \in X_i \cap (\omega_i + \sum_{j\in J} \theta_{i,j} Y_j + Z)$ is an autarky consumption (Assumption 1.(d)) for consumer i. Let $s \in ([0,1] \cap \mathbb{Q})^I$ be such that $t_i = 0 \Longrightarrow s_i = 0$ and $t_i > 0 \Longrightarrow 1 - \varepsilon < \frac{t_i}{s_i} \leq 1$.

Set $x_{i,s} = \frac{t_i}{s_i} x_{i,t} + (1 - \frac{t_i}{s_i})\underline{x}_i$ for each i such that $t_i > 0$, and notice that $x_{i,s} \in P_i(x_i)$. Using (5.1), a simple calculation shows that

$$\sum_{\{i\in I:\ s_i>0\}} s_i x_{i,s} - \sum_{i\in I} s_i \omega_i \in \sum_{\{i\in I:\ s_i>0\}} s_i \frac{t_i}{s_i} \sum_{j\in J} \theta_{i,j} Y_j$$

$$+ \sum_{\{i\in I:\ s_i>0\}} s_i\left(1 - \frac{t_i}{s_i}\right) \sum_{j\in J} \theta_{i,j} Y_j + Z \subset \sum_{i\in I} s_i \sum_{j\in J} \theta_{i,j} Y_j + Z$$

by convexity of the Y_j. This contradicts $x \in \mathcal{C}^e(\mathcal{E})$. □

Before closing this section, we should emphasize that the fact that Assumptions 1, 2, 3, 4 parallel the ones used in the finite dimensional case does not mean that these assumptions are easily satisfied and mild from the point of view of the economic interpretation.

Recall that with the definition of the economy $\mathcal{E}$ is given a commodity–price duality $\langle L, L' \rangle$, which means that we shall look for quasiequilibrium prices belonging to L'. Let τ be a vector space topology compatible with this duality. For applying Proposition 5.2.3, the main difficulty is in the choice of a Hausdorff vector space topology σ on L such that the set $\widehat{X}$ of all attainable consumption allocations be σ^I-compact. In the next two propositions, we give (strong) sufficient conditions for the τ-boundedness[6] of each attainable consumption set. We then indicate in corollaries how to choose σ so as to get σ-compactness of each attainable consumption set. In the first proposition, the commodity space (L, τ) is an ordered topological vector space whose topology τ is such that order intervals (and, consequently, all order-bounded subsets of L) are τ-bounded. In the second one, the commodity space is a normed vector lattice.[7]

[6] Recall that a subset A of a topological vector space is τ-bounded if for every V τ-neighborhood of 0, there exists $\lambda > 0$ such that $A \subset \lambda V$.

[7] A normed vector lattice is a vector lattice equipped with a monotone norm, that is, with a norm verifying: $|x| \leq |y| \Longrightarrow \|x\| \leq \|y\|$.

PROPOSITION 5.2.4 (ALIPRANTIS–BROWN–BURKINSHAW [7]) *Assume that the commodity space (L, τ) is an ordered topological vector space whose topology τ is such that order intervals are τ-bounded. Consider the conditions:*

a *Each consumption set X_i, $i \in I$, is order-bounded from below by some $a_i \in L$;*

b *The total production set Y is order-bounded from above by some $b \in L$;*

c *Each production set Y_j, $j \in J$, is order-bounded from above by some $b_j \in L$.*

Under conditions a and b, each attainable consumption set $\widehat{X}_i$ is order-bounded, thus τ-bounded. If condition b is replaced by the stronger condition c, then each attainable production set $\widehat{Y}_j$ is also order-bounded, thus τ-bounded.

Proof. Recalling that Z is order-bounded from above by 0, the proof of the first assertion is immediate if we notice that for each $x_i \in \widehat{X}_i$, there are some $y \in Y$, $z \in Z$, and $x_k \in X_k$ for each $k \neq i$ such that

$$a_i \leq x_i = \omega + y + z - \sum_{k \neq i} x_k \leq \omega + b - \sum_{k \neq i} a_k.$$

Let us now assume a and c. That c implies b is obvious. Fix $j \in J$. If $y_j \in \widehat{Y}_j$, let $x \in \prod_{i \in I}$, $(y_{j'})_{j' \neq j} \in \prod_{j' \neq j} Y_j$ and $z \in Z$ be such that $\sum_{i \in I} x_i = \omega + y_j + \sum_{j' \neq j} y_{j'} + z$. From $a \leq \sum_{i \in I} x_i = \omega + y_j + + \sum_{j' \neq j} y_{j'} + z$ we deduce

$$a - \sum_{j' \neq j} b_{j'} - \omega \leq a - \sum_{j' \neq j} y_{j'} - z - \omega \leq y_j \leq b_j,$$

which proves that $\widehat{Y}_j$ is order-bounded, thus τ-bounded. □

PROPOSITION 5.2.5 (BEWLEY [35]) *Assume that the commodity space (L, τ) is a normed vector lattice. Consider the conditions:*

a *For each $i \in I$, the consumption set X_i is order-bounded from below by some $a_i \in L$;*

b *If $a = \sum_{i \in I} a_i$, the set $(\omega + Y + Z) \cap (a + L_+)$ is norm-bounded;*

c *If we set* $Z = Y_0$, *then for each* $j \in J \cup \{0\}$, *for each* $u \in L$, *the set* $Y_j \cap (L_+ - \sum_{j' \neq j} Y_{j'} + u)$ *is norm-bounded.*

Under conditions a and b, each attainable consumption set $\widehat{X}_i$ *is norm-bounded. Under conditions a and c, each attainable consumption set* $\widehat{X}_i$ *and each attainable production set* $\widehat{Y}_j$, $j \in J \cup \{0\}$ *is norm-bounded.*

Proof. Under condition a, if $x \in \widehat{X}$, then $\sum_{i \in I} x_i \in (\omega + Y + Z) \cap (a + L_+)$. Under condition b, this set is norm-bounded. Since the norm of L is monotone, the relation $0 \leq x_i - a_i \leq \sum_{i \in I} x_i - a$ implies that each $\widehat{X}_i$ is norm-bounded.

Let us now assume a and c and fix $j \in J \cup \{0\}$. If $y_j \in \widehat{Y}_j$, let $x \in \prod_{i \in I} X_i$ and $(y_{j'})_{j' \neq j} \in \prod_{j' \neq j} Y_j$ be such that $\sum_{i \in I} x_i = \omega + y_j + \sum_{j' \neq j} y_{j'}$. This shows that $y_j \in L_+ - \sum_{j' \neq j} Y_{j'} + a - \omega$. It follows from condition c that $\widehat{Y}_j$ is norm-bounded. If $x \in \widehat{X}$ then $\sum_{i \in I} x_i \in \omega + \sum_{j \in J \cup \{0\}} \widehat{Y}_j$. As previously, since $\sum_{i \in I} x_i$ belongs to a norm-bounded set and since the norm is monotone, each $\widehat{X}_i$ is norm-bounded. □

Looking for a topology σ on the commodity space L such that the set $\widehat{X}$ of all attainable consumption allocations be σ^I-compact, let us assume either that the topology τ on L is such that order-intervals are τ-compact or that L is the topological dual of a barrelled topological vector space M and that τ is the strong topology for this duality. In the latter case, τ-bounded subsets of L are $\sigma(L, M)$-bounded and it follows from the Alaoglu–Bourbaki theorem[8] that they are relatively $\sigma(L, M)$-compact. Taking $\sigma = \tau$ in the first case and $\sigma = \sigma(L, M)$ in the second case, Proposition 5.2.4 has the following corollary.

COROLLARY 5.2.1 *In the conditions of Proposition 5.2.4, under one of the two previous conditions on the topology* τ, *and with the appropriate choice of* σ, *assume that all consumption sets* X_i *and* $Y + Z$ *are* σ*-closed. Then under a and b, the set* $\widehat{X}$ *of all attainable consumption allocations is* σ^I*-compact. Assume that all consumption sets* X_i *and all production sets* Y_j, $j \in J$ *are* σ*-closed. Then, under conditions a and c, the set* $\widehat{X}$ *is* σ^I*-compact and each attainable production set is* σ*-compact.*

Proof. The verification of σ-closedness of the relevant attainable sets is immediate and left for the reader. □

[8]See [44, Chap.4,§2, Theorem 1].

Assume now that L is the dual of a Banach lattice M and that the lattice norm given on L is the dual norm of the norm on M.[9] Taking for τ the norm topology on L and recalling that τ-bounded subsets of L are $\sigma(L, M)$-bounded and thus relatively $\sigma(L, M)$-compact, we have the following corollary to Proposition 5.2.5.

COROLLARY 5.2.2 *In the conditions of Proposition 5.2.5, assuming that L is the dual of a Banach lattice M, taking $\sigma = \sigma(L, M)$, and letting $\tau(L, M)$ denote the Mackey topology for the duality $\langle L, M\rangle$, assume that all consumption sets X_i and $Y + Z$ are $\tau(L, M)$-closed. Then under a and b, the set $\widehat{X}$ of all attainable consumption allocations is σ^I-compact. If we assume that all consumption sets X_i, all production sets Y_j, and Z are $\tau(L, M)$-closed, then, under conditions a and c, the set $\widehat{X}$ is σ^I-compact and each attainable production set is σ-compact.*

Proof. It suffices to notice that, being convex sets, the sets X_i, $i \in I$, Y_j, $j \in J$ and $Y + Z$ are $\sigma(L, M)$-closed if they are assumed to be $\tau(L, M)$-closed. □

The previous topological assumptions fit well with all the benchmark examples of commodity spaces that have been given in the first section of this chapter. The choice of the topology σ determines the strength of the continuity condition set on preferences in Assumption 1(b). When the commodity space is some $L = L^{\ell}_{\infty}(\Omega, \Sigma, \mu)$ and $\sigma = \sigma(L_\infty, L_1)$, Assumption 1(b) is a strong assumption, interpreted as a condition of asymptotic impatience (see [18], [35],[50],[152]). When $L = M(K)$ and $\sigma = \sigma(M(K), C(K))$, Assumption 1(b) is standard, once it has been accepted that "close" for individual consumption bundle means $\sigma(M(K), C(K))$-close. As explained in Jones [109], continuity of preferences with respect to this topology is verified in standard location models and in the characteristics model of Lancaster [121] and [122].

A last remark is in order.

REMARK 5.2.1 Assume that $\mathcal{E} = (\langle L, L'\rangle, (X_i, u_i, \omega_i)_{i\in I})$ is an exchange economy without disposal, and that consumers' preferences are derived from utility functions $u_i: X_i \to \mathbb{R}$, normalized by requiring $u_i(\omega_i) = 0$. Let

$$\mathcal{U} = \{v = (v_i)_{i\in I} \in \mathbb{R}^I_+ \text{: for some } x \in \widehat{X}, \text{ for each } i \in I, 0 \le v_i \le u_i(x_i)\}$$

[9]Recall that a Banach lattice is a complete normed vector lattice. If the lattice norm given on L is the dual norm of the norm on M, then L is also a Banach lattice. For a discussion about vector lattices, see page 123.

be the set of attainable and individually rational utility vectors, simply called *individually rational utility set.* In Condition 1 stated above for the nonemptiness of the fuzzy core of $\mathcal{E}$, the σ^I-compactness of $\widehat{X}$ can be replaced by the compactness of $\mathcal{U}$. More precisely, the existence of Edgeworth equilibria for $\mathcal{E}$ is proved in [17] under the condition:

1′ $\mathcal{U}$ is compact. Moreover for each $i \in I$,

- (a) X_i is convex and $\omega_i \in X_i$,
- (b) $u_i: X_i \to \mathbb{R}$ is quasi-concave.

Adding the above condition 4 stated page 115, the fuzzy core $\mathcal{C}^f(\mathcal{E})$ is nonempty. Condition 4 is in particular satisfied if the functions u_i are concave. Condition 4 also satisfied if the functions u_i are τ-lower semi-continuous at every attainable and individually rational consumption vector.

To assume compactness of $\mathcal{U}$ instead of σ^I-compactness of $\widehat{X}$ allows to drop the assumption that individual consumption sets are order-bounded from below. This is especially useful in finance where short-selling of assets is allowed and where compactness of $\mathcal{U}$ is interpreted as a *nonarbitrage* condition on the financial economy $\mathcal{E}$. For more details on this assumption and sufficient conditions on the primitives of $\mathcal{E}$ for this condition to be fulfilled, see [17] and its list of references.

5.3 Decentralizing Edgeworth allocations

By decentralizing Edgeworth allocations, we mean associating with an Edgeworth allocation prices in L' for which this allocation is a quasiequilibrium allocation. The techniques for obtaining decentralizing prices differ very much according to the properties of the compatible topology we consider on the commodity space. When an interiority assumption to be precised later is possible, the result and its proof parallel the statement and the proof of Proposition 4.3.4 in Chapter 4. Without interiority assumptions, we will assume that the commodity space is a Riesz space and will introduce the so-called properness assumptions on the characteristics of the economy.

5.3.1 Decentralization under interiority assumptions

Let τ be a topology on L compatible with the duality $\langle L, L' \rangle$. We start with an attainable consumption allocation $\overline{x} \in \mathcal{C}^f(\mathcal{E})$.

PROPOSITION 5.3.1 *Let $\overline{x}$ be an element of the fuzzy core of $\mathcal{E}$ and let $\overline{y} \in \prod_{j\in J} Y_j$, $\overline{z} \in Z$ be such that $\sum_{i\in I} \overline{x}_i - \omega - \sum_{j\in J} \overline{y}_j = \overline{z}$. Assume that either Z or the set $\bigcup_{i\in I}(P_i(\overline{x}) - \sum_{j\in J} \theta_{i,j} Y_j - \omega_i)$ has a nonempty τ-interior. Then there is a nonzero price system $p \in L'$ such that $(\overline{x}, \overline{y}, p)$ is a quasiequilibrium of $\mathcal{E}$, provided the following conditions are satisfied:*

1. *for each $i \in I$, the preferred set $P_i(\overline{x})$ is convex and $\overline{x}_i \in \operatorname{cl} P_i(\overline{x})$, the τ-closure of $P_i(\overline{x})$,*
2. *for each $j \in J$, the production set Y_j is convex.*

The quasiequilibrium is nontrivial if $\omega \in \operatorname{int}(\sum_{i\in I} X_i - Y - Z)$, the τ-interior of $(\sum_{i\in I} X_i - Y - Z)$.

Proof. Define $G = \operatorname{co}\Big(\bigcup_{i\in I}(P_i(\overline{x}) - \sum_{j\in J} \theta_{i,j} Y_j - \omega_i)\Big)$. We first claim that the set G is nonempty and that $G \cap Z = \emptyset$. Indeed, by Conditions (1) and (2), each $P_i(\overline{x}) - \sum_{j\in J} \theta_{i,j} Y_j - \omega_i$ is nonempty and convex. So, by an elementary convexity result,

$$G = \Big\{\sum_{i\in I} \lambda_i (x_i - \sum_{j\in J} \theta_{i,j} y_{i,j} - \omega_i) \colon x_i \in P_i(\overline{x}),\ y_{i,j} \in Y_j,\ \lambda_i \geq 0,\ \sum_{i\in I} \lambda_i = 1\Big\}.$$

Exactly as in the proof of Proposition 4.3.4 one verifies that $\overline{x} \in \mathcal{C}^f(\mathcal{E})$ implies $G \cap Z = \emptyset$. Moreover, by assumption, either Z has a nonempty interior, or the set G, being the convex hull of a set with a nonempty τ-interior, has also a nonempty τ-interior.

We thus can apply the Hahn–Banach separation theorem for disjoint convex sets of which one has a nonempty interior. Let $p \neq 0$ in $L' = (L, \tau)'$, the topological dual of (L, τ), and $\alpha \in \mathbb{R}$ be such that

$$p \cdot Z \leq \alpha \leq p \cdot G.$$

the remainder of the proof is routine. Since Z is a cone, $\alpha \geq 0$ and $p \cdot Z \leq 0 \leq p \cdot G$. From this, we deduce:

$$x_i \in P_i(\overline{x}) \Longrightarrow p \cdot x_i \geq p \cdot \omega_i + \sum_{j\in J} \theta_{i,j} p \cdot Y_j. \tag{5.3}$$

Using Condition (1), we have for each $i \in I$:

$$p \cdot \overline{x}_i \geq p \cdot \omega_i + \sum_{j\in J} \theta_{i,j} p \cdot Y_j, \text{ and, in particular,} \tag{5.4}$$

$$p \cdot \overline{x}_i \geq p \cdot \omega_i + \sum_{j \in J} \theta_{i,j} p \cdot \overline{y}_j. \tag{5.5}$$

Summing over $i \in I$, we get $p \cdot \sum_{i \in I}(\overline{x}_i - \omega_i) \geq p \cdot \sum_{j \in J} \overline{y}_j$. From $\sum_{i \in I}(\overline{x}_i - \omega_i) - \sum_{j \in J} \overline{y}_j \in Z$, we deduce $p \cdot \sum_{i \in I}(\overline{x}_i - \omega_i) = p \cdot \sum_{j \in J} \overline{y}_j$, and, finally, for each $i \in I$,

$$p \cdot \overline{x}_i = p \cdot \omega_i + \sum_{j \in J} \theta_{i,j} p \cdot \overline{y}_j. \tag{5.6}$$

Using (5.3) and (5.6) we get:

$$x_i \in P_i(\overline{x}) \Longrightarrow p \cdot x_i \geq p \cdot \overline{x}_i = p \cdot \omega_i + \sum_{j \in J} \theta_{i,j} p \cdot \overline{y}_j.$$

Let now $j \in J$ and $y_j \in Y_j$. Using (5.4) and (5.6) for i such that $\theta_{i,j} > 0$, we get $p \cdot y_j \leq p \cdot \overline{y}_j$. We have thus proved

$$y_j \in Y_j \Longrightarrow p \cdot y_j \leq p \cdot \overline{y}_j.$$

The two last implications, with $p \in Z^0$ and

$$p \cdot \Big(\sum_{i \in I} (\overline{x}_i - \omega_i) - \sum_{j \in J} \overline{y}_j \Big) = 0,$$

complete the proof.

The proof of the last assertion is straightforward. □

REMARK 5.3.1 Clearly, to assume in the previous proposition that the disposal cone Z has a nonempty τ-interior restricts the application of this decentralization result to commodity spaces (L, τ) whose positive cone has a nonempty interior. It is under this hypothesis that were obtained, with an approach different of the one used in this chapter, the free disposal equilibrium existence results of Bewley([34] and [35]).[10]

An interiority assumption made on some $P_i(\overline{x})$ meets the same objection when one looks at the order boundedness assumptions that insure

[10]Bewley [35] demonstrates equilibria in finite dimensional sub-economies and establishes a net of these equilibria that converges to an equilibrium on the whole commodity space. When extending positive prices in the finite dimensional economy to positive linear functionals on the whole commodity space, this approach uses also explicitly the nonemptiness of the norm-interior of the positive cone of $L_\infty(\Omega, \Sigma, \mu)$, the space of essentially bounded measurable functions defined on a σ-finite positive measure space (Ω, Σ, μ).

via Propositions 5.2.4 and 5.2.5 the existence of attainable consumption allocations in the fuzzy core. If consumption sets are to be bounded from below, an interiority assumption on some $P_i(\overline{x})$ implies that the positive cone of the commodity space has a nonempty interior.

An interiority assumption made as in Duffie [65] on some "augmented" total production set yielding the same attainable allocations as the original total production set is a severe assumption which may be useful in particular examples.

In order to get decentralization results in commodity spaces whose positive cone may have an empty interior, the so-called properness assumptions compensate for the lack of interiority of preferred and production sets. Before introducing the various properness concepts which will be used in the sequel, we first comment the assumptions on the commodity–price duality which, associated with properness, will allow to get quasiequilibrium existence.

5.3.2 An alternative assumption on the commodity–price duality

From now on, we make the following structural assumption on the commodity–price duality:

[**SA**] L is a linear vector lattice (or Riesz space) equipped with a Hausdorff locally convex linear topology τ such that

a The positive cone L_+ is closed for the τ-topology of L;

b $L' = (L, \tau)'$, the topological dual of (L, τ), is a vector sublattice of the order dual $L^\sim$ of L.

Assumption [SA] combines several ingredients of which each one needs some comments.

- First, with the assumption that the commodity space is a linear vector lattice, it is assumed that any two elements x and y of L have a supremum denoted by $x \vee y$ and an infimum denoted by $x \wedge y$. This allows to see each element of L as the difference $x = x^+ - x^-$ of two positive elements, namely its *positive component* $x^+ = x \vee 0$ and its *negative component* $x^- = -(x \wedge 0) = (-x) \vee 0$. As a consequence of this property, the positive cone L_+ is *generating.* The absolute value $|x|$ of x is defined as $|x| = x^+ + x^-$.

  Vector lattices or Riesz spaces have two main properties which are extensively used in the literature devoted to the existence of general equilibrium in vector lattices. The first one is proved by elementary calculations.

THEOREM 5.3.1 (RIESZ DECOMPOSITION PROPERTY) *Let* x, y_1, y_2 *in* L_+ *satisfy* $0 \leq x \leq y_1 + y_2$. *Then there exist elements* x_1, x_2 *of* L *such that* $0 \leq x_1 \leq y_1$, $0 \leq x_2 \leq y_2$ *and* $x = x_1 + x_2$.

By induction, one immediately deduces that if x and finitely many other elements of L_+ satisfy $0 \leq x \leq \sum_{i=1}^{n} y_i$ then there exist elements $x_1, \ldots, x_n$ such that $0 \leq x_i \leq y_i$ for each i, and $x = \sum_{i=1}^{n} x_i$.

Before stating the second property, let us recall some notations. If $a, b \in L$ satisfy $a \leq b$, the *order interval* $[a, b]$ is by definition

$$[a, b] := \{x \in L\colon\ a \leq x \leq b\}.$$

The vector space of all linear functionals f on L such that the image by f of any order interval is an order bounded subset of $\mathbb{R}$ is called *order dual* of L and denoted by $L^{\sim}$. The order dual $L^{\sim}$ can be ordered by the relation: $f \geq g$ if $f(x) \geq g(x)$ for every $x \in L_+$. Moreover, we have the following.

THEOREM 5.3.2 (RIESZ–KANTOROVICH) *If* L *is an ordered vector space with the Riesz decomposition property, then the order dual* $L^{\sim}$ *is a Riesz space and for each* $f, g \in L^{\sim}$ *and* $x \in L_+$ *its lattice operations are given by*

1 $[f \vee g](x) = \sup\{f(y) + g(z)\colon\ y, z \in L_+ \ \text{ and } \ y + z = x\}$, *and*

2 $[f \wedge g](x) = \inf\{f(y) + g(z)\colon\ y, z \in L_+ \ \text{ and } \ y + z = x\}$.

In particular, note that if L has the Riesz decomposition property, then for any finite collection of linear functionals $f_1, f_2, \ldots, f_m \in L^{\sim}$ their supremum in $L^{\sim}$ at each $x \in L_+$ is given by

$$\Big[\bigvee_{i=1}^{m} f_i\Big](x) = \sup\Big\{\sum_{i=1}^{m} f_i(x_i)\colon\ x_i \in L_+ \text{ for each } i \text{ and } \sum_{i=1}^{m} x_i = x\Big\}. \tag{$\star$}$$

For $f = (f_1, \ldots, f_m)$, the formula ($\star$) that gives the supremum of the functionals $f_1, \ldots, f_m$ is called the *Riesz–Kantorovich formula* of these functionals and for $x \geq 0$ in L, $\Big[\bigvee_{i=1}^{m} f_i\Big](x)$ is denoted by $\mathcal{R}_f(x)$.

Vector subspaces K of L such that $x \in K$, and $|y| \leq |x|$ imply $y \in K$ will play a particular role in the sequel. A vector subspace K satisfying this property is called *order ideal* of L. Using the formulas which combine vector and lattice operations of L, it may be seen that an order ideal is a *vector sublattice* of L, that is, for any couple of elements of K the supremum in L of x and y belongs to K and consequently coincides with the supremum in K of x and y. A vector sublattice of L needs not be an order ideal of L. For more details regarding Riesz spaces we refer the reader to [8]. Some notions or results will be recalled when used in the text.

- It is next assumed in [SA] that the vector space topology τ considered on L is Hausdorff and locally convex. As well-known, Hausdorff locally convex topological vector spaces have important properties. In particular, $\langle L, (L, \tau)' \rangle$ is a duality and a Mackey theorem states that topologically bounded subsets are the same for every Hausdorff locally convex topology compatible with this duality. Among other ones this property will be used in the sequel.

- The required compatibility between the order structure and the topological structure of the commodity space is expressed by conditions (a) and (b) of [SA]. Condition (a) states a kind of minimal compatibility which makes the ordered vector space L to be what is usually called an "ordered topological vector space". Condition (b) states that the topological dual $(L, \tau)'$ is not only a vector subspace[11] but a vector sublattice of the order dual $L^{\sim}$, that is, any finite collection of continuous linear functionals on L has a supremum in L' the value of which can be calculated using the Riesz–Kantorovich formula.

The reader will verify (for more explanations, see [7], [8], [132]) that conditions (a) and (b) are automatically satisfied if, as in Mas-Colell [130], the commodity space L is a Hausdorff locally convex topological vector lattice and if L' is the topological dual of L. Commodity–prices dualities as $\langle L_p, L_q \rangle$ with $1 \leq p < \infty$, $\frac{1}{p} + \frac{1}{q} = 1$, or the commodity–price dualities formed by each of L_∞ or $M(K)$ with their respective norm-dual satisfy Mas-Colell's assumptions. In counterpart, Assumption [SA] with (a)

[11] Together with the assumption that τ is a Hausdorff locally convex topology, the inclusion of $(L, \tau)'$ in $L^{\sim}$ implies that order intervals of L are topologically bounded.

and (b) allows for considering commodity–price dualities as $\langle L_\infty, L_1\rangle$ or $\langle M(K), C(K)\rangle$ with any compatible topology on the commodity space.[12]

5.3.3 Decentralization under uniform properness in a Hausdorff locally convex topological vector lattice

In this subsection, we assume that the commodity space (L, τ) is a topological vector lattice, that is, τ has a local base of locally convex-solid neighborhoods of 0,[13] and that the consumption set of each agent coincides with the positive cone L_+. Recall that $L' = (L, \tau)'$.

For preferences, the original properness definitions are given as follows for an economy defined in a Hausdorff locally convex ordered topological vector commodity space.

DEFINITION 5.3.1 (MAS-COLELL) *Let $\succsim$ be a complete preorder on L_+. The preference relation $\succsim$ is said to be* ***pointwise proper*** *at $x \in L_+$ if there exist $v > 0$ and a neighborhood U of 0 such that if $\lambda > 0$ and $x - \lambda v + \lambda u \succsim x$ then $u \notin U$.*

The preference relation $\succsim$ is ***uniformly proper*** *if it is pointwise-proper at every $x \in L_+$ with v and U independent of x.*

DEFINITION 5.3.2 (YANNELIS–ZAME) *Let $P\colon L_+ \to L_+$ be a preference correspondence. The correspondence P is said to be* ***pointwise F-proper*** *at $x \in L_+$ if there exist $v > 0$ and a neighborhood U of 0 such that if $\lambda > 0$, $u \in U$ and $x + \lambda v + \lambda u \in L_+$, then $x + \lambda v + \lambda u \in P(x)$.*

The correspondence P is ***uniformly F-proper*** *if it is pointwise F-proper at every $x \in L_+$ with v and U independent of x.*

The geometrical difference (see Figure 5.1) between the two previous definitions is obvious. Mas-Colell's pointwise properness at x of a complete preorder requires the existence of a cone Γ_x (with vertex 0), with a nonempty interior, such that the set $x - \Gamma_x$ lie below the indifference set of x. If P is the strict preference correspondence associated with a complete preorder, pointwise F-properness at x (F is for forward) requires the existence of a cone Γ_x (with vertex 0), with a nonempty interior, such that the set $x + \Gamma_x$ point inward the indifference set of x and verify

[12]Notice however that L_∞ equipped with the Mackey topology $\tau(L_\infty, L_1)$ is a Hausdorff locally convex topological vector lattice.

[13]For a 0-neighborhood $U \in \mathcal{B}(0)$ of this base, U is convex and $[u \in U$ and $|s| \le |u| \Longrightarrow s \in U]$.

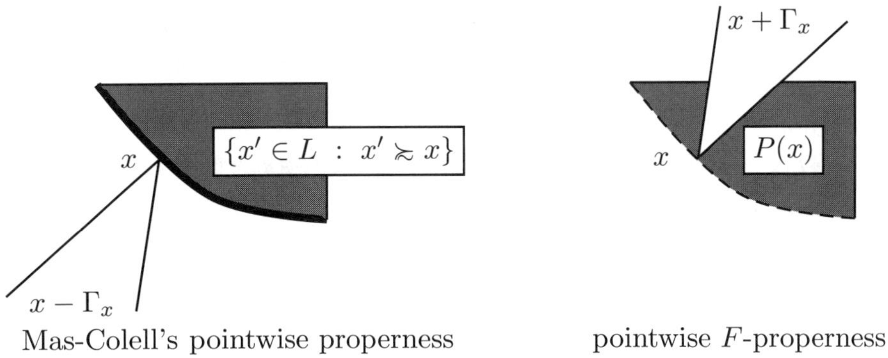

Figure 5.1.

in addition $(x+\Gamma_x) \cap L_+ \subset P(x)$. When $P: L_+ \to L_+$ is the strict preference associated with a complete preorder $\succsim$, it is easy to see that uniform F-properness of P coincides with Mas-Colell's uniform properness of the complete preorder $\succsim$. For this reason, uniform F-properness is considered as the nontransitive version of uniform properness as defined by Mas-Colell. However, even when P is the strict preference correspondence associated with a complete preorder, pointwise properness of the preorder and pointwise F-properness of the associated strict preference correspondence are incomparable.

On the production side, assuming that L is a vector lattice equipped with a Hausdorff locally convex vector space topology, pointwise and uniform properness are defined as follows.

DEFINITION 5.3.3 (RICHARD) *Let $Y \subset L$ be a production set. The set Y is said to be* ***pointwise F-proper*** *at $y \in Y$ if there exist $v > 0$ and a neighborhood U of 0 such that if $\lambda > 0$, $u \in U$ and $(y - \lambda v + \lambda u)^+ \leq y^+$ (equivalently, $\lambda u^+ \leq y^- + \lambda v$) then $y - \lambda v + \lambda u \in Y$.*

The set Y is ***uniformly F-proper*** *if it is F-pointwise-proper at every $y \in Y$ with v and U independent of x.*

The F-properness of Y at $y \in Y$ means that it is possible to substitute any other "small" commodity bundle for an additional increment in the opposite direction to the direction of properness and still get a feasible production, provided that the obtained output is less than or equal to the initial output. Uniform F-properness means that this is "uniformly" possible at any point of the production set.

Mas-Colell [131] gives a stronger definition of uniform properness.

DEFINITION 5.3.4 (MAS-COLELL) *The set Y is **uniformly proper** if there exist $v > 0$, a neighborhood U of 0 and a closed convex sublattice Z of L containing Y and 0 and satisfying $Z - L_+ = Z$, such that if $\lambda > 0$, $u \in U$ and $y - \lambda v + \lambda u \in Z$ then $y - \lambda v + \lambda u \in Y$. The closed, convex and comprehensive lattice Z is called **pretechnology set**.*

To see that uniform properness is stronger than uniform F-properness, observe that Z as in Definition 5.3.4 contains all lattices $Z_y = \{z \in L\colon z^+ \leq y^+\}$ for $y \in Y$. Indeed, for each $y \in Y$, $y^+ \in Z$ and $z^+ \leq y^+$ implies $z^+ \in Z$ and thus $z \in Z$ since $z \leq z^+$. It is worth noticing that the free-disposal production set $Y = -L_+$ satisfies uniform F-properness, but that $Y = \{0\}$ does not satisfies F-properness.

DEFINITION 5.3.5 *The economy*

$$\mathcal{E} = \Big(\langle L, L' \rangle, (L_+, P_i, \omega_i)_{i \in I}, (Y_j)_{j \in J}, (\theta_{i,j})_{\substack{i \in I \\ j \in J}}, Z\Big)$$

*is **uniformly F-proper** if each correspondence P_i is uniformly F-proper and the production set $T := \sum_{j \in J} Y_j + Z$ is uniformly F-proper.*

REMARK 5.3.2 In the previous definitions, without loss of generality, the properness neighborhoods can be assumed convex and solid. Using the decomposition property of vector lattices and solidness of properness neighborhoods, the reader will verify that if each production set Y_j is uniformly F-proper with $(v_j,\, U_j)$ as properness constants and if the disposal cone Z is either reduced to $\{0\}$ or uniformly F-proper at z with $(v_Z,\, U_Z)$ as properness constants, then $T = \sum_{j \in J} Y_j + Z$ is uniformly F-proper with either $(v_T = \sum_{j \in J} v_j,\, U_T = \bigcap_{j \in J} U_j)$ or $(v_T = \sum_{j \in J} v_j + v_z$, $U_T = (\bigcap_{j \in J} U_j) \cap U_Z)$ as properness constants. It follows that an economy whose preference correspondences, production sets and, if case, the disposal cone are uniformly F-proper is a uniformly F-proper economy.

Assume now that $\mathcal{E}$ is a uniformly F-proper economy, and set $v = \sum_{i \in I} v_i + v_T$ and $U = (\bigcap_{i \in I} U_i) \cap U_T$, where all v_i and v_T are the properness vectors and all properness neighborhoods U_i and U_T are assumed to be convex and solid. Let

$$\Gamma := \{z \in L\colon\ z = \lambda(-v + u),\ \lambda \geq 0,\ u \in U\}$$

denote the convex cone with vertex 0 generated by $-v + U$. Let also $P\colon X \to X$ denote the unanimous preference correspondence defined by

$$P(x) = \{x' \in X\colon\ x'_i \in P_i(x) \text{ for each } i \in I\}.$$

Recall that $\widehat{X} = \{x \in \prod_{i\in I} X_i : \sum_{i\in I} x_i \in \omega + Y + Z\}$. We first prove a useful result.

PROPOSITION 5.3.2 *Under the previous conditions, if* $x \in \prod_{i\in I} X_i$ *satisfies* $\sum_{i\in I} x_i \in \omega + Y + Z + \Gamma$ *then* $\widehat{X} \cap (P(x) \cup \{x\}) \neq \emptyset$.

Proof. Assume that for $x \in \prod_{i\in I} X_i$, we have $\sum_{i\in I} x_i \in \omega + Y + Z + \Gamma$, that is

$$\sum_{i\in I} x_i = \omega + t - \lambda(\sum_{i\in I} v_i + v_T) + \lambda u \tag{5.7}$$

for some $\lambda \geq 0$, $t \in T = Y + Z$, $u \in (V_T \cap (\bigcap_{i\in I} V_i))$.

If $\lambda = 0$, there is nothing to prove.

If $\lambda > 0$, we deduce from (5.7) that $\lambda u \leq \sum_{i\in I}(x_i + \lambda v_i) + (t^- + \lambda v_T)$, thus that

$$\lambda u^+ \leq \sum_{i\in I}(x_i + \lambda v_i) + (t^- + \lambda v_T). \tag{5.8}$$

Using the decomposition property, let $(s_i)_{i\in I}$ and s_T be such that

$$\lambda u^+ = \sum_{i\in I} \lambda s_i + \lambda s_T$$

with for each i, $0 \leq \lambda s_i \leq (x_i + \lambda v_i)$, and $0 \leq \lambda s_T \leq (t^- + \lambda v_T)$. Replacing in (5.7) λu^+ by its value, we get

$$\sum_{i\in I}(x_i + \lambda v_i - \lambda s_i) = \omega + t - \lambda v_T + \lambda(s_T - u^-). \tag{5.9}$$

We now observe that in view of solidness of U, for each i, $s_i \in U \subset U_i$ which, together with the positivity of $x_i + \lambda v_i - \lambda s_i$ and the uniform F-properness of P_i, implies that $x_i + \lambda v_i - \lambda s_i \in P_i(x)$. On the other hand, since $s_T \geq 0$,

$$|s_T - u^-| \leq s_T \vee u^- \leq u^+ \vee u^- \leq |u| \tag{5.10}$$

and it follows from solidness of the properness neighborhoods that $(s_T - u^-) \in U \subset U_T$. Together with $\lambda s_T \leq (t^- + \lambda v_T)$ and the uniform F-properness of T, this implies $\lambda(s_T - u^-)^+ \leq t^- + \lambda v_T$, then $t - \lambda v_T + \lambda(s_T - u^-) \in T$.

We have thus proved that if $\lambda > 0$ then $(x_i + \lambda v_i - s_i)_{i\in I} \in \widehat{X} \cap P(x)$. □

Let us now associate with the uniformly F-proper economy $\mathcal{E}$ the auxiliary economy

$$\mathcal{E}_\Gamma = \left(\langle L, L'\rangle, (X_i, P_i, \omega_i)_{i\in I}, (Y_j)_{j\in J}, (\theta_{i,j})_{\substack{i\in I\\ j\in J}}, Z, \Gamma\right)$$

deduced from $\mathcal{E}$ by adding a fictitious producer whose production set is the cone Γ defined as above.[14] Using Proposition 5.3.1, we will decentralize an element of the fuzzy core of $\mathcal{E}_\Gamma$. We first give sufficient conditions for nonemptiness of $\mathcal{C}^f(\mathcal{E}_\Gamma)$.

PROPOSITION 5.3.3 *Using the above notations, assume that τ has a local base $\mathcal{B}(0)$ of σ-closed locally convex-solid neighborhoods of* 0.[15] *If in addition to the conditions 1, 2, 3, 4 for the nonemptiness of $\mathcal{C}^f(\mathcal{E})$, each X_i, Y_j and Z are σ-closed and each $\widehat{Y}_j$ is σ-compact, then $\mathcal{C}^f(\mathcal{E}_\Gamma) \neq \emptyset$.*

Proof. Without loss of generality, we can assume that the properness neighborhoods are σ-closed, convex and solid. We first observe that it follows from the irreflexivity of strict preference correspondences and the solidness of properness neighborhoods that $v \notin U$ and consequently the cone Γ is σ-closed.

Let $\mathcal{K}$ be the collection of the convex and σ-compact subsets K of L containing 0, each $\widehat{X}_i$, each $\widehat{Y}_j$, $\widehat{Z}$ and, in view of Assumption 1(d), all $\underline{x}_i$, $\underline{y}_{i,j}$,$\underline{z}_i$ such that $\omega_i = \underline{x}_i - \sum_{j\in J}\theta_{i,j}\underline{y}_{i,j} - \underline{z}_i$. For each $K \in \mathcal{K}$, if $x \in \prod_{i\in I}(X_i \cap K)$, we set $P_{i,K}(x) = P_i(x) \cap K$ and we consider the economy

$$\mathcal{E}_{\Gamma,K} = \left(\langle L, L'\rangle, (X_i \cap K, P_{i,K}, \omega_i)_{i\in I}, (Y_j \cap K)_{j\in J}, (\theta_{i,j})_{\substack{i\in I\\ j\in J}}, Z \cap K, \Gamma\right).$$

We next observe that the set $\widehat{X}_{\Gamma,K}$ of all attainable consumption allocations of this economy is σ^I-compact. Indeed, the set

$$\mathcal{A}(\mathcal{E}_{\Gamma,K}) = \left\{(x,y,z,\gamma) \in \prod_{i\in I}(X_i \cap K) \times \prod_{j\in J}(Y_j \cap K) \times (Z \cap K) \times \Gamma : \right.$$
$$\left. \sum_{i\in I} x_i = \omega + \sum_{j\in J} y_j + z + \gamma \right\}$$

[14]Since Γ is a cone with vertex 0, it is not necessary to precise profit shares for consumers.

[15]It is for instance the case if L is the dual of a locally convex-solid vector lattice and if τ is the strong topology for this duality.

is σ-closed, thus σ-compact. Being the projection of $\mathcal{A}(\mathcal{E}_{\Gamma,K})$ on the set $\prod_{i\in I}(X_i \cap K)$, $\widehat{X}_{\Gamma,K}$ is σ-compact.

It then follows from Proposition 5.2.3 that $\mathcal{C}^f(\mathcal{E}_{\Gamma,K}) \neq \emptyset$.[16] Let $\overline{x}_K \in \mathcal{C}^f(\mathcal{E}_{\Gamma,K})$. Using the definition of $\mathcal{K}$, it follows from Proposition 5.3.2 that $\overline{x}_K \in \widehat{X}$.

Now let us remark that the collection $\mathcal{K}$ ordered by inclusion is directed. Since $\widehat{X}$ is σ^I-compact, by passing to subnets if necessary, we can assume that $\overline{x}_K \xrightarrow{\sigma^I} \overline{x} \in \widehat{X}$. We claim that $\overline{x} \in \mathcal{C}^f(\mathcal{E}_\Gamma)$. Indeed, if $\overline{x} \notin \mathcal{C}^f(\mathcal{E}_\Gamma)$, it is easy to see that there is some $K_0 \in \mathcal{K}$ such that $K \supset K_0 \Longrightarrow \overline{x}_K \notin \mathcal{C}^f(\mathcal{E}_{\Gamma,K})$. This yields a contradiction, proving the claim and completing the proof of the proposition. □

COROLLARY 5.3.1 *Under the previous assumptions, the economy $\mathcal{E}$ has a quasiequilibrium $(\overline{x}, \overline{y}, \overline{p})$ such that $\overline{p} \cdot v > 0$. This quasiequilibrium is non trivial if for some $\lambda > 0$ in $\mathbb{R}$, $\lambda v \in \omega + \sum_{j\in J} Y_j + Z - \sum_{i\in I} X_i$.*

Proof. According to the previous proposition, let $\overline{x} \in \mathcal{C}^f(\mathcal{E}_\Gamma)$ and let $\overline{y}$ be such that $\sum_{i\in I} \overline{x}_i - \omega - \sum_{j\in J} \overline{y}_j \in Z$. In view of Proposition 5.3.1, there is $\overline{p} \in L' = (L, \tau)'$ such that $(\overline{x}, \overline{y}, \overline{p})$ is a quasiequilibrium of $\mathcal{E}_\Gamma$. Since the allocation $(\overline{x}, \overline{y})$ is attainable for $\mathcal{E}$, then $(\overline{x}, \overline{y}, \overline{p})$ is actually a quasiequilibrium of $\mathcal{E}$. Moreover since Γ is a cone and since $-v \in \operatorname{int}\Gamma$, then $\overline{p} \cdot \Gamma \leq 0 \Longrightarrow \overline{p} \cdot v > 0$. Consequently, if $\lambda v \in \omega + \sum_{j\in J} Y_j + Z - \sum_{i\in I} X_i$, then the quasiequilibrium $(\overline{x}, \overline{y}, \overline{p})$ is nontrivial. □

REMARK 5.3.3 Assume that ω is the common properness direction for each P_i and T, for instance that $v_T = v_i = \frac{\omega}{|I|+1}$ for each $i \in I$. The condition for nontriviality of the quasiequilibrium is in particular satisfied if $0 \in \sum_{j\in J} Y_j$. Indeed from $\omega \in \sum_{i\in I} X_i - \sum_{j\in J} Y_j - Z$, that is, $-\omega \in \sum_{j\in J} Y_j + Z - \sum_{i\in I} X_i$, and $0 \in \sum_{j\in J} Y_j + Z - \sum_{i\in I} X_i$, one deduces by convexity that for every $\lambda:\ 0 < \lambda < 1$, one has $(1-\lambda)(-\omega) \in \sum_{j\in J} Y_j + Z - \sum_{i\in I} X_i$, that is, $\lambda\omega \in \omega + \sum_{j\in J} Y_j + Z - \sum_{i\in I} X_i$. The nontriviality condition is consequently satisfied in exchange economies where consumption sets coincide with the positive cone and, under the same condition, in production economies where inaction is possible, at least for the total production set.

Corollary 5.3.1 extends to nontransitive preferences the quasiequilibrium existence results of Mas-Colell [130] and Richard [153]. It extends Yan-

[16] The set $Z \cap K$ is not a cone. However, a cautious reader could observe that in the proof of Proposition 5.2.3, the assumption that Z is a cone is superfluous.

nelis and Zame [181] to topological vector lattice commodity spaces and to production economies. It extends also our result [82] to the case where there is no disposal. The comparison of Richard's uniform properness assumptions for production used in this subsection with Zame's [182] assumption of a bounded *marginal efficiency of the total production* is more difficult.

5.3.4 Decentralization under properness

We now turn to the full generality of the assumptions on the commodity–price duality stated in Subsection 5.3.1 and do no longer assume that consumption sets of the agents coincide with the positive cone of the commodity space.

Let us assume $\omega > 0$. The properness assumptions we will use in this subsection for preferred sets and production sets are now borrowed from Tourky ([171] and [172]). They are formulated with respect to the topology τ used in the statement of Assumption [SA].

DEFINITION 5.3.6 *P_i is ω-**proper** at the ith component of $x \in \prod_{i \in I} X_i$ if $x_i \in \operatorname{cl} P_i(x)$ (the τ-closure of $P_i(x)$) and if there are a convex set $\widehat{P}_i(x) \subset L$ and a lattice $Z_{x_i} \subset L$ such that*

1 $\widehat{P}_i(x) \cap Z_{x_i} = P_i(x)$;

2 $x_i + \omega$ *is a τ-interior point of* $\widehat{P}_i(x)$;

3 $x_i,\ 0,\ \omega_i \in Z_{x_i}$, *and* $Z_{x_i} + L_+ \subset Z_{x_i}$.

Note that it follows from 3 (more precisely, from $Z_{x_i} + L_+ \subset Z_{x_i}$) that, in this definition, the lattice Z_{x_i} is convex.

DEFINITION 5.3.7 *Y_j is ω-**proper** at $y_j \in Y_j$ if there are a convex set $\widehat{Y}_j(y_j) \subset L$ and a lattice $Z_{y_j} \subset L$ such that*

1 $\widehat{Y}_j(y_j) \cap Z_{y_j} = Y_j$;

2 $y_j - \omega$ *is a τ-interior point of* $\widehat{Y}_j(y_j)$;

3 $0 \in Z_{y_j}$, *and* $Z_{y_j} - L_+ \subset Z_{y_j}$.

Once again, it follows from 3 (more precisely, from $Z_{y_j} - L_+ \subset Z_{y_j}$) that, in this definition, the lattice Z_{y_j} is convex.

If the disposal cone $Z \neq \{0\}$, we consider this set as a production set and make on Z an analogue properness assumption.

DEFINITION 5.3.8 *Z is ω-**proper** at $z \in Z$ if there are a convex set $\widehat{Z}(z) \subset L$ and a lattice $Z_z \subset L$ such that*

1 $\widehat{Z}(z) \cap Z_z = Z$;

2 $z - \omega$ is a τ-interior point of $\widehat{Z}(z)$;

3 $0 \in Z_z$, and $Z_z - L_+ \subset Z_z$.

DEFINITION 5.3.9 *The economy*

$$\mathcal{E} = \Big(\langle L, L' \rangle, (X_i, P_i, \omega_i)_{i \in I}, (Y_j)_{j \in J}, (\theta_{i,j})_{\substack{i \in I \\ j \in J}}, Z \Big)$$

*is ω-**proper** at $(x, y, z) \in \prod_{i \in I} X_i \times \prod_{j \in J} Y_j \times Z$ if each $P_i(x)$ is ω-proper at x_i, each Y_j is ω-proper at y_j and the disposal cone Z is either reduced to $\{0\}$ or ω-proper at z.*

REMARK 5.3.4 If, in Definition 5.3.6, the positive cone L_+ of the space is both the consumption set of consumer i and the lattice Z_{x_i} used to state properness at $x_i \in L_+$ of his preference correspondence, the properness assumption takes a more familiar guise since it states that P_i is proper at $x_i \in \operatorname{cl} P_i(x)$ if the preferred set $P_i(x) \subset L_+$ is extendible to some convex set with a nonempty interior. In this case, ω-properness is a stronger assumption than a slightly weaker form of pointwise F-properness defined by Podczeck [149] with ω as properness vector. A slightly weaker form of properness is defined in Podczeck [149] and called E-properness. In case of a complete preference preordering on L_+, for each $x_i \in L_+$, both assumptions of properness at x_i are implied by Mas-Colell's [130] original definition of uniformly ω-proper preferences and are proved to be strictly weaker.[17]

To see that a uniformly proper complete preference preordering $\succsim_i$ on the positive cone is proper at each point of the positive cone, assume that preferences $\succsim_i$ are uniformly proper on L_+ and let U_i be the properness neighborhood, and Γ_i be the cone generated by $(\omega + U_i)$. If P_i denotes the strict preference correspondence associated with $\succsim_i$, it follows from transitivity and completeness of the preorder $\succsim_i$ and from uniformity of Mas-Colell's properness that $(P(x_i) + \Gamma_i) \cap L_+ \subset P_i(x_i)$ holds for each $x_i \in L_+$. Then $\widehat{P}_i(x_i) = P_i(x_i) + (\{0\} + \Gamma_i)$ satisfies at x_i the conditions of Definition 5.3.6.

[17]See [149] and [171].

But let us emphasize that for nonordered preferences defined on the positive cone, uniform F-properness and properness are incomparable assumptions.

Properness at y_j as in Definition 5.3.7 is implied by Mas-Colell's [131] original definition of uniformly ω-proper production,[18] used by Richard in [154]. Indeed, let U_j be the properness neighborhood, Γ_j be the cone generated by $(-\omega + U_j)$ and Z_j be the pretechnology set such that $(y_j + \Gamma_j) \cap Z_j \subset Y_j$ holds for each $y_j \in Y_j$. Without loss of generality, U_j and consequently Γ_j can be assumed to be convex. Then $\widehat{Y}_j = Y_j + (\{0\} + \Gamma_j)$ satisfies for every $y_j \in Y_j$ the conditions of Definition 5.3.7.

A third observation is in order. A disposal cone Z equal to the negative cone $-L_+$ is ω-proper at any $z \in Z$. Indeed, $\widehat{Y}_0(z) := L$ and $Z_z := -L_+$ satisfy the condition of Definition 5.3.8.

Properness assumptions and extension of linear functionals. The main tool of the extension result which will be proved in the next proposition is a convexity result inspired by Podczeck [149, Lemma 2].

LEMMA 5.3.1 (PODCZECK) *Let (L, τ) be a Hausdorff topological vector space and let K be a vector subspace of L. Let A be a convex subset of K and V be a convex τ-open subset of L such that $V \cap A \neq \emptyset$. If p is a linear functional on K satisfying for some $a \in A \cap \operatorname{cl} V$*

$$p \cdot a \leq p \cdot x \text{ for each } x \in V \cap A \tag{5.11}$$

then there exists a τ-continuous linear functional $\pi \in (L, \tau)'$ and a linear functional h on L such that $p = \pi_{|K} + h_{|K}$ and

$$\pi \cdot a \leq \pi \cdot x \text{ for each } x \in V, \quad h \cdot a \leq h \cdot x \text{ for each } x \in A. \tag{5.12}$$

Let us assume in addition that (L, τ) is an ordered vector space and set $K_+ = L_+ \cap K$. If $A + K_+ \subset A$ then $h_{|K} \geq 0$, $\pi_{|K} \leq p$, and

$$p \cdot (a - x) = \pi \cdot (a - x) \text{ for each } x \leq a, \ x \in A. \tag{5.13}$$

Proof. Consider the set

$$T = \{(x', r) \in L \times \mathbb{R} \colon \exists x \in A, \ v \in V \text{ such that } x' = v - x \text{ and } r > p(x)\}.$$

[18] which should not be confused with uniform F-properness as stated in Definition 5.3.3.

Using the assumptions on A and V, it is easily seen that the set T is a nonempty convex subset of $L \times \mathbb{R}$, open if T is equipped with the product of the topologies on L and $\mathbb{R}$. It follows from (5.11) that $(0, p \cdot a) \notin T$. Thus from the Hahn–Banach theorem and the openness of T there exists a nonzero $(\pi, \lambda) \in (L, \tau)' \times \mathbb{R}$ such that

$$\pi \cdot x' + \lambda r > \lambda p \cdot a$$

for all $x' = v - x$, $v \in V$, $x \in A$, and $r > p \cdot x$. In particular, if $x \in A \cap V$ and $r > p(x)$, it follows from (5.11) that $r > p(a)$ thus that $\lambda > 0$. Without lost of generality one can set $\lambda = 1$. We thus have

$$\pi \cdot v - \pi \cdot x + p \cdot x \geq p \cdot a$$

for all $v \in V$ and $x \in A$. Applied to $x = a$, this relation yields $\pi \cdot a \leq \pi \cdot x$ for each $x \in V$. On the other hand, from $a \in \operatorname{cl} V$ we deduce $(p - \pi) \cdot x \geq (p - \pi) \cdot a$ for each $x \in A$. Let $h \colon L \to \mathbb{R}$ be any linear functional extending $p - \pi_{|K}$. One has $p = \pi_{|K} + h_{|K}$ and π and h satisfy (5.12).

To prove the second assertion, let us assume in addition that (L, τ) is an ordered vector space, that K is ordered by the induced order ($K_+ = L_+ \cap K$) and that $A + K_+ \subset A$. From $h \cdot x \geq h \cdot a$ for each $x \in A$ we deduce $h \cdot (a + K_+) \geq h \cdot a$, thus $h \cdot K_+ \geq 0$. On the other hand, if $x \leq a$, $x \in A$ then $h \cdot x \geq h \cdot a$ by (5.12) and $h \cdot x \leq h \cdot a$ by $h \cdot K_+ \geq 0$, thus $h \cdot x = h \cdot a$ and $p \cdot (a - x) = \pi \cdot (a - x)$ as required in (5.13). □

Replacing the condition $A + K_+ \subset A$ by $A - K_+ \subset A$ and applying the previous lemma to the sets $-V$, $-A$ and to the point $-a$, we get immediately the following corollary.

COROLLARY 5.3.2 *Let (L, τ) be a Hausdorff topological vector space and let K be a vector subspace of L. Let A be a convex subset of K and V be a convex τ-open subset of L such that $V \cap A \neq \emptyset$. If p is a linear functional on K satisfying for some $a \in A \cap \operatorname{cl} V$*

$$p \cdot a \geq p \cdot y \text{ for each } y \in V \cap A \tag{5.14}$$

then there exists a τ-continuous linear functional $\pi \in (L, \tau)'$ and a linear functional h on L such that $p = \pi_{|K} + h_{|K}$ and

$$\pi \cdot a \geq \pi \cdot y \text{ for each } y \in V, \quad h \cdot a \geq h \cdot y \text{ for each } y \in A. \tag{5.15}$$

Let us assume in addition that (L, τ) is an ordered vector space and that $K_+ = L_+ \cap K$. If $A - K_+ \subset A$ then $h_{|K} \geq 0$, $\pi_{|K} \leq p$, and

$$p \cdot (a - y) = \pi \cdot (a - y) \text{ for each } y \geq a, \ y \in A. \tag{5.16}$$

Exploiting the previous lemma and its corollary, the following proposition is the main result of this paragraph. In what follows, if K is a vector subspace of L, the truncated economy $\mathcal{E}_{|_K}$ is defined from $\mathcal{E}$ by letting $X_{i,K} = X_i \cap K$, $Y_{j,K} = Y_j \cap K$, $Z_K = Z \cap K$, and if $x \in \prod_{k \in I} X_{i,k}$, $P_{i,K}(x) = P_i(x) \cap K$.

PROPOSITION 5.3.4 *Let K be some order ideal of L containing all ω_i and let $(\overline{x}, \overline{y}, \overline{p})$ be a quasiequilibrium of $\mathcal{E}_{|_K}$ with a price $\overline{p} \in K^{\sim}$, the order dual of K. Let $\overline{z} \in Z_K$ be such that $\sum_{i \in I} \overline{x}_i = \sum_{i \in I} \omega_i + \sum_{j \in J} \overline{y}_j + \overline{z}$. If the economy $\mathcal{E}$ is ω-proper at $(\overline{x}, \overline{y}, \overline{z})$, then there exist τ-continuous linear functionals π_i, $i \in I$, π_j, $j \in J$, and π_0 if $Z \neq \{0\}$, such that, with the notations of Definitions 5.3.6, 5.3.7, 5.3.8,*

$$\pi_i \cdot \widehat{P}_i(\overline{x}) \geq \pi_i \cdot \overline{x}_i, \quad \pi_j \cdot \widehat{Y}_j(\overline{y}_j) \leq \pi_j \cdot \overline{y}_j, \quad i \in i, \ j \in J, \tag{5.17}$$

$$\pi_0 \cdot \widehat{Z}(\overline{z}) \leq \pi_0 \cdot \overline{z} = 0 \quad \text{if } Z \neq \{0\}. \tag{5.18}$$

Moreover, let π be the collection $((\pi_i)_{i \in I}, (\pi_j)_{j \in J}, \pi_0)$ if $Z \neq \{0\}$ (resp. $((\pi_i)_{i \in I}, (\pi_j)_{j \in J})$ if $Z = \{0\}$) and $\overline{\pi} = \mathcal{R}_{\pi}$. Then $\overline{\pi} \in (L, \tau)'$ and

$$\forall i \in I, \forall x_i \leq \overline{x}_i, \ x_i \in Z_{\overline{x}_i} \cap K, \ \pi_i \cdot (\overline{x}_i - x_i) = \overline{\pi} \cdot (\overline{x}_i - x_i) = \overline{p} \cdot (\overline{x}_i - x_i), \tag{5.19}$$

$$\forall j \in J, \forall y_j \geq \overline{y}_j, \ y_j \in Z_{\overline{y}_j} \cap K, \ \pi_j \cdot (\overline{y}_j - y_j) = \overline{\pi} \cdot (\overline{y}_j - y_j) = \overline{p} \cdot (\overline{y}_j - y_j), \tag{5.20}$$

$$\forall z \geq \overline{z}, \ z \in Z_{\overline{z}} \cap K, \ \pi_0 \cdot (\overline{z} - z) = \overline{\pi} \cdot (\overline{z} - z) = \overline{p} \cdot (\overline{z} - z), \ \text{if } Z \neq \{0\}, \tag{5.21}$$

$$\overline{\pi} \cdot (\omega - u) = \overline{p} \cdot (\omega - u) \text{ for each } u \leq \omega, u \in \sum_{i \in I} Z_{\overline{x}_i} \cap K - \sum_{j \in J} Z_{\overline{y}_j} \cap K - Z_{\overline{z}} \cap K, \tag{5.22}$$

and finally, for every $i \in I$

$$\overline{\pi} \cdot \overline{x}_i = \overline{\pi} \cdot \omega_i + \sum_{j \in J} \theta_{i,j} \overline{\pi} \cdot \overline{y}_j. \tag{5.23}$$

Proof. Since $(\overline{x}, \overline{y}, \overline{p})$ is a quasiequilibrium of $\mathcal{E}_{|_K}$ and the economy $\mathcal{E}$ is ω-proper at $(\overline{x}, \overline{y}, \overline{z})$, letting $\text{int}(\widehat{P}_i(\overline{x}))$ denote the τ-interior of $\widehat{P}_i(\overline{x})$, we have for every $i \in I$

$$\overline{x}_i + \omega \in \text{int}(\widehat{P}_i(\overline{x})) \cap (Z_{\overline{x}_i} \cap K)$$

$$\overline{x}_i \in \operatorname{cl} P_i(\overline{x}) \cap Z_{\overline{x}_i} \cap K \subset \operatorname{cl}\Big(\operatorname{int}(\widehat{P}_i(\overline{x}))\Big) \cap (Z_{\overline{x}_i} \cap K),$$

$$p \cdot (\widehat{P}_i(\overline{x}) \cap (Z_{\overline{x}_i} \cap K)) = p \cdot ((\widehat{P}_i(\overline{x}) \cap Z_{\overline{x}_i}) \cap K) \geq p \cdot \overline{x}_i.$$

Applying the first part of Lemma 5.3.1 with $a = \overline{x}_i$, $A = Z_{\overline{x}_i} \cap K$ and $V = \operatorname{int}(\widehat{P}_i(\overline{x}))$, we get for each i the existence of $\pi_i \in (L, \tau)'$ such that $\pi_i \cdot \operatorname{int}(\widehat{P}_i(\overline{x})) \geq \pi_i \cdot \overline{x}_i$ which implies

$$\pi_i \cdot \widehat{P}_i(\overline{x}) \geq \pi_i \cdot \overline{x}_i.$$

Since $Z_{\overline{x}_i} + L_+ \subset Z_{\overline{x}_i}$, we get from the second part of Lemma 5.3.1 that $\pi_{i|_K} \leq \overline{p}$ and

$$(\overline{p} - \pi_i) \cdot \overline{x}_i = (\overline{p} - \pi_i) \cdot x_i \quad \text{for each } x_i \leq \overline{x}_i,\ x_i \in Z_{\overline{x}_i} \cap K. \tag{5.24}$$

Using symmetrical arguments and Corollary 5.3.2, one gets for each j the existence of $\pi_j \in (L, \tau)'$ such that $\pi_{j|_K} \leq \overline{p}$ and

$$\pi_j \cdot \widehat{Y}_j(\overline{y}_j) \leq \pi_j \cdot \overline{y}_j,$$

$$(\overline{p} - \pi_j) \cdot \overline{y}_j = (\overline{p} - \pi_j) \cdot y_j \quad \text{for each } y_j \geq \overline{y}_j,\ y_j \in Z_{\overline{y}_j} \cap K. \tag{5.25}$$

Likewise, if $Z \neq \{0\}$ then there exists $\pi_0 \in (L, \tau)'$ such that $\pi_{0|_K} \leq \overline{p}$ and

$$\pi_0 \cdot \widehat{Z}(\overline{z}) \leq \pi_0 \cdot \overline{z},$$

$$(\overline{p} - \pi_0) \cdot \overline{z} = (\overline{p} - \pi_0) \cdot z \quad \text{for each } z \geq \overline{z},\ z \in Z_{\overline{z}} \cap K. \tag{5.26}$$

From $0 \geq \overline{z}$ and $0 \in Z_{\overline{z}}$ we deduce $\pi_0 \cdot \overline{z} = \overline{p} \cdot \overline{z} = 0$, which ends the proof of (5.17) and (5.18).

Now, it follows from the structural assumption [SA] on the commodity–price duality that $\overline{\pi}$, as defined in the proposition, is τ-continuous, that is, belongs to L'. Moreover, let us recall that K is an order ideal of L, thus that $\overline{\pi}$ coincide on K with the supremum of all $\pi_{i|_K}$, $\pi_{j|_K}$, and $\pi_{0|_K}$ if $Z \neq \{0\}$. So $\overline{\pi}_{|_K} \leq p$. The assertions (5.19), (5.20), (5.21) follow from this remark and respectively from (5.24), (5.25), (5.26). To prove (5.22), assume that $u = \sum_{i \in I} u_i - \sum_{j \in J} u_j - u_0 \leq \omega$ with $u_i \in Z_{\overline{x}_i} \cap K$, $u_j \in Z_{\overline{y}_j} \cap K$ and $u_0 = 0$ or $u_0 \in Z_{\overline{z}} \cap K$ if $Z \neq \{0\}$. Define for each i, $z_i = u_i \wedge \overline{x}_i$, for each j, $z_j = u_j \vee \overline{y}_j$, and $z_0 = u_0 \vee \overline{y}_j$. Since the sets $Z_{\overline{x}_i}$, $Z_{\overline{y}_j}$, $Z_{\overline{z}}$ are lattices containing respectively $\overline{x}_i$, $\overline{y}_j$, $\overline{z}$ and since K is an order ideal of L containing all $\overline{x}_i$, $\overline{y}_j$, and $\overline{z}$, we have $z_i \in Z_{\overline{x}_i} \cap K$, $z_j \in Z_{\overline{y}_j} \cap K$, $z_0 \in Z_{\overline{z}} \cap K$ and

$$0 \leq \omega - u \leq \sum_{i \in I} (\overline{x}_i - z_i) + \sum_{j \in J} (z_j - \overline{y}_j) + (z_0 - \overline{z}).$$

From $\overline{\pi}_{|K} \leq p$, we deduce $0 \leq (p - \overline{\pi}) \cdot (\omega - u)$. On the other hand, $(p - \overline{\pi}) \cdot (\omega - u) \leq (p - \overline{\pi}) \cdot (\sum_{i \in I}(\overline{x}_i - z_i) + \sum_{j \in J}(z_j - \overline{y}_j) + (z_0 - \overline{z}))$ and it follows from (5.19), (5.20), (5.21) that $(p - \overline{\pi}) \cdot (\omega - u) \leq 0$.

Finally, for each $i \in I$, let $z_i = \overline{x}_i \wedge \omega_i$. Since $\overline{x}_i$ and $\omega_i \in Z_{\overline{x}_i} \cap K$, $Z_{\overline{x}_i}$ is a lattice and K an order ideal of L, $z_i \in Z_{\overline{x}_i} \cap K$. It thus follows from (5.19) that $\overline{\pi} \cdot \overline{x}_i = p \cdot \overline{x}_i + (\overline{\pi} - p) \cdot z_i$. Recalling that $\pi_{i|K} \leq p$, we get

$$\overline{\pi} \cdot \overline{x}_i \geq p \cdot \overline{x}_i + (\overline{\pi} - p) \cdot \omega_i$$

Since $(\overline{x}, \overline{y}, \overline{p})$ is a quasiequilibrium of $\mathcal{E}_{|K}$, we deduce

$$\overline{\pi} \cdot \overline{x}_i \geq \overline{\pi} \cdot \omega_i + \sum_{j \in J} \theta_{i,j} p \cdot \overline{y}_j \quad \text{for each } i \in I. \tag{5.27}$$

Now, for each $j \in J$, $\overline{y}_j$ and $0 \in Z_{\overline{y}_j} \cap K$, $Z_{\overline{y}_j}$ is a lattice and K an order ideal of L. One thus has $\overline{y}_j^+ \in Z_{\overline{y}_j} \cap K$ and it follows from (5.20) that $\overline{\pi} \cdot (\overline{y}_j^+ - \overline{y}_j) = p \cdot (\overline{y}_j^+ - \overline{y}_j)$ and thus that $p \cdot \overline{y}_j = \overline{\pi} \cdot \overline{y}_j + (p - \overline{\pi}) \cdot \overline{y}_j$. Using again $\pi_{i|K} \leq p$, we get

$$p \cdot \overline{y}_j \geq \overline{\pi} \cdot \overline{y}_j \quad \text{for each } j \in J \tag{5.28}$$

and, combining (5.27) with (5.28),

$$\overline{\pi} \cdot \overline{x}_i \geq \overline{\pi} \cdot \omega_i + \sum_{j \in J} \theta_{i,j} \overline{\pi} \cdot \overline{y}_j \quad \text{for each } i \in I.$$

That all these inequalities are actually equalities comes from the fact that $\sum_{i \in I} \overline{x}_i - \sum_{j \in J} \overline{y}_j - \omega = \overline{z}$ and $\overline{\pi} \cdot \overline{z} = 0$. □

COROLLARY 5.3.3 *If, in the conditions of Proposition 5.3.4, $\overline{p} \cdot \omega > 0$, then $(\overline{x}, \overline{y}, \overline{z}, \overline{\pi}_{|K})$ is a quasiequilibrium of $\mathcal{E}_{|K}$ such that $\overline{\pi} \cdot \omega > 0$.*

Proof. It suffices to verify that the price $\overline{\pi}_{|K}$ supports the allocation $(\overline{x}, \overline{y}, \overline{z})$ in the economy $\mathcal{E}_{|K}$.

Indeed, let $x_i \in P_i(\overline{x}) \cap K = \widehat{P}_i(\overline{x}) \cap Z_{\overline{x}_i} \cap K$ and $z_i = x_i \wedge \overline{x}_i$. Since K and $Z_{\overline{x}_i}$ are respectively an order ideal and a lattice containing both x_i and $\overline{x}_i$, $z_i \in Z_{\overline{x}_i} \cap K$ and it follows from (5.17), (5.19) and the definition of $\overline{\pi}$ that

$$\overline{\pi} \cdot (x_i - z_i) \geq \pi_i \cdot (x_i - z_i) \geq \pi_i \cdot (\overline{x}_i - z_i) = \overline{\pi} \cdot (\overline{x}_i - z_i),$$

thus that $\overline{\pi} \cdot x_i \geq \overline{\pi} \cdot \overline{x}_i$.

Let $y_j \in Y_j \cap K = \widehat{Y}_j \cap Z_{\overline{y}_j} \cap K$ and $z_j = y_j \vee \overline{y}_j$. Since K and $Z_{\overline{y}_j}$ are respectively an order ideal and a lattice containing both y_j and $\overline{y}_j$,

$z_j \in Z_{\overline{y}_j} \cap K$ and it follows from (5.17), (5.20) and the definition of $\overline{\pi}$ that

$$\overline{\pi} \cdot (y_j - z_j) \leq \pi_j \cdot (y_j - z_j) \leq \pi_j \cdot (\overline{y}_j - z_j) = \overline{\pi} \cdot (\overline{y}_j - z_j),$$

thus that $\overline{\pi} \cdot y_j \leq \overline{\pi} \cdot \overline{y}_j$.

If $Z \neq 0$, let $z \in Z \cap K = Z \cap \widehat{Z}_{\overline{z}}$ and $z_0 = z \vee \overline{z}$. Since K and $Z_{\overline{z}}$ are respectively an order ideal and a lattice containing both z and $\overline{z}$, $z_0 \in Z_{\overline{z}} \cap K$ and it follows from (5.18), (5.21) and the definition of $\overline{\pi}$ that

$$\overline{\pi} \cdot (z - z_0) \leq \pi_0 \cdot (z - z_0) \leq \pi_0 \cdot (\overline{z} - z_0) = \overline{\pi} \cdot (\overline{z} - z_0),$$

thus that $\overline{\pi} \cdot z \leq \overline{\pi} \cdot \overline{z}$. Since $Z \cap K$ is a cone, $\overline{\pi} \cdot \overline{z} = 0$.

It follows from (5.22) that $\overline{\pi} \cdot \omega = p \cdot \omega > 0$ and from the relations (5.23) that $(\overline{x}, \overline{y}, \overline{z}, \overline{\pi}_{|K})$ is a quasiequilibrium of $\mathcal{E}_{|K}$. □

The interest of the previous corollary which will not be used in the sequel is to show that a quasiequilibrium allocation $(\overline{x}, \overline{y}, \overline{z})$ supported by a quasiequilibrium price $p \in L^{\sim}$ such that $p \cdot \omega > 0$ can also be supported by a τ-continuous quasiequilibrium price, provided that the economy $\mathcal{E}$ is ω-proper at $(\overline{x}, \overline{y}, \overline{z})$.

Application to the quasiequilibrium existence problem. As in Subsection 5.3.1, we start from an attainable consumption allocation $\overline{x} \in \mathcal{C}^f(\mathcal{E})$ and define $\overline{y} = (\overline{y}_j)_{j\in J} \in \prod_{j\in J} Y_j$ and $\overline{z} \in Z$ such that $\sum_{i\in I} \overline{x}_i = \sum_{i\in I} \omega_i + \sum_{j\in J} \overline{y}_j + \overline{z}$.

Let K be any principal order ideal in L generated by some $u > 0$ and containing all $\overline{x}_i$, $\overline{y}_j$, ω_i. We first observe that if $\mathcal{E}_{|K}$ denotes the truncated economy as previously defined, then $\overline{x} \in \mathcal{C}^f(\mathcal{E}_{|K})$. Assuming that $\omega > 0$ and $\mathcal{E}$ is ω-proper at $(\overline{x}, \overline{y}, \overline{z})$ and using Proposition 5.3.1, we will first prove that the allocation $(\overline{x}, \overline{y})$ can be decentralized as a quasiequilibrium of $\mathcal{E}_{|K}$. More precisely,

PROPOSITION 5.3.5 *Let us assume that $\omega > 0$ and that $\mathcal{E}$ is ω-proper at $(\overline{x}, \overline{y}, \overline{z})$. Then there exists a price system $\overline{p} \in K^{\sim}$ such that $\overline{p} \cdot \omega > 0$ and $(\overline{x}, \overline{y}, \overline{p})$ is a quasiequilibrium of $\mathcal{E}_{|K}$.*

Proof. Let $K = \{v \in L\colon\ |v| \leq \lambda u \text{ for some } \lambda > 0\}$. Notice that when K is equipped with the Riesz norm

$$\|v\|_u = \inf\{\lambda > 0\colon\ |v| \leq \lambda u\},$$

the order intervals $[-\lambda u, +\lambda u]$, $\lambda > 0$ form a basis for the 0-neighborhoods of this topology, thus u is an interior point of the positive cone K_+. Moreover, on K the Riesz norm topology is finer than the topology induced by the topology τ considered on L. Indeed, it follows from [SA] (b) (see footnote 11) that the order interval $[-u, +u]$ is τ-bounded, thus that for every τ-neighborhood V of 0 in L there is some $\lambda > 0$ such that $[-u, +u] \subset \lambda V$. Together with the assumption that K is an order ideal of L, this implies that $V \cap K$ is a neighborhood of 0 for the Riesz norm topology of K.

From this observation and the properness of each $P_i(\overline{x})$ at $\overline{x}_i$, we deduce successively that for each $i \in I$, any $0 < \alpha \leq 1$, and some τ-neighborhood V of 0 ,

$$\overline{x}_i + \alpha\omega + V \subset \widehat{P}_i(\overline{x}),$$

$$\overline{x}_i + \alpha\omega + (V \cap K) \subset (\overline{x}_i + \alpha\omega + V) \cap K \subset P_i(\overline{x}) \cap K,$$

and for some $\lambda > 0$,

$$\overline{x}_i + \alpha\omega + \frac{1}{\lambda}[-u, +u] \subset P_i(\overline{x}),$$

that is, $\overline{x}_i + \alpha\omega$ is an interior point of $P_i(\overline{x}) \cap K$ for the Riesz norm topology.

This shows that the set $\bigcup_{i\in I}\Big((P_i(\overline{x}) \cap K) - \sum_{j\in J} \theta_{i,j}(Y_j \cap K) - \omega_i\Big)$ has a nonempty Riesz norm-interior. This shows also that for each $i \in I$, $\overline{x}_i$ belongs to the Riesz norm closure of $P_i(\overline{x}) \cap K$. Applying Proposition 5.3.1 in Subsection 5.3.1, we get the existence of a nonzero price system $\overline{p} \in (K, \|\cdot\|_u)'$ such that $(\overline{x}, \overline{y}, \overline{p})$ is a quasiequilibrium of $\mathcal{E}_{|K}$.

It remains to verify that $\overline{p} \in L^\sim$ and that $\overline{p} \cdot \omega > 0$. Notice that, by construction, the Riesz norm on K is monotone ($|v| \leq |w| \Longrightarrow \|v\|_u \leq \|w\|_u$) and makes of K a normed Riesz space. That the topological dual is a vector subspace of the order dual is a classical property of normed Riesz spaces and more generally of Hausdorff locally convex topological vector lattices.[19] That $\overline{p} \cdot \omega > 0$ follows from properness, for example from the fact that for each i, $\overline{x}_i + \omega$ is an interior point of $P_i(\overline{x}) \cap K$ for the Riesz norm topology. □

To go further, we now consider the family $\mathcal{K}$ of all principal order ideals K of L containing all $\overline{x}_i$, $\overline{y}_j$, ω_i and observe that $\mathcal{K}$ is a directed family

[19] Regarding the details of the proof, the reader will benefit of reading Aliprantis–Burkinshaw [8, Chapter 2] and Schaefer [161, Chapter V 6.2].

for the set inclusion. In order to decentralize $(\overline{x}, \overline{y})$ as a quasiequilibrium of the original economy, we will now apply the previous proposition and the extension result of the previous paragraph to each order ideal $K \in \mathcal{K}$, and then will pass to limit.

The proof of the next proposition will strongly rely on an observation stated in the next lemma which extends inductively to any finite collection of nets of order bounded linear functional L.

Let $\overline{\mathbb{R}} = \mathbb{R} \cup \{-\infty\} \cup \{+\infty\}$, topologized in the usual way. Recall first that if $\mathcal{K}$ is a directed set and if $(x^K)_{K \in \mathcal{K}}$ is a net of elements of $\overline{\mathbb{R}}$, $\limsup_K x^K := \inf_K \sup_{K' \geq K} x^{K'}$ and $\liminf_K x^K := \sup_K \inf_{K' \geq K} x^K$ exist in $\overline{\mathbb{R}}$ and verify $\liminf_K x^K \leq \limsup_K x^K$. If moreover $\lim_K x^K$ exists in $\overline{\mathbb{R}}$, then $\liminf_K x^K = \limsup_K x^K = \lim_K x^K$.

LEMMA 5.3.2 *Let $\mathcal{K}$ be a directed set and let $(\pi_1^K)_{K \in \mathcal{K}}$, $(\pi_2^K)_{K \in \mathcal{K}}$ be two nets indexed on $\mathcal{K}$ of order bounded linear functionals on L. Let for each $K \in \mathcal{K}$, $\pi^K = \pi_1^K \vee \pi_2^K$ and assume that the nets $(\pi_1^K)_{K \in \mathcal{K}}$ and $(\pi_2^K)_{K \in \mathcal{K}}$ converge respectively to $\pi_1 \in L^\sim$ and $\pi_2 \in L^\sim$ for the weak-star topology $\sigma(L^\sim, L)$. Then if $\pi = \pi_1 \vee \pi_2$, we have for every $a \in L_+$*

$$\liminf_K \pi^K \cdot a \geq \pi \cdot a\,.$$

Proof. Let $a \in L_+$. For each K and for every $a_1 \geq 0$, $a_2 \geq 0$ in L such that $a_1 + a_2 = a$, it follows from the definition of π^K that $\pi_1^K \cdot a_1 + \pi_2^K \cdot a_2 \leq \pi^K \cdot a$. Passing to limit, one gets $\pi_1 \cdot a_1 + \pi_2 \cdot a_2 \leq \liminf_K \pi^K \cdot a$. Since the previous relation holds for every $a_1 \geq 0$, $a_2 \geq 0$ in L such that $a_1 + a_2 = a$, it follows from the definition of π that $\pi \cdot a \leq \liminf_K \pi^K \cdot a$. □

PROPOSITION 5.3.6 *Let us assume that $\omega > 0$ and that $\mathcal{E}$ is ω-proper at $(\overline{x}, \overline{y}, \overline{z})$. Then there exists a price system $\overline{\pi} \in L'$ such that $\overline{\pi} \cdot \omega > 0$ and $(\overline{x}, \overline{y}, \overline{\pi})$ is a quasiequilibrium of $\mathcal{E}$. This quasiequilibrium is nontrivial if for some $\lambda > 0$ in $\mathbb{R}$, $\lambda\omega \in \omega + \sum_{j \in J} Y_j + Z - \sum_{i \in I} X_i$.*

Proof. For each $K \in \mathcal{K}$, in view of the previous proposition, let $\overline{p}^K \in K^\sim$ be such that $\overline{p}^K \cdot \omega > 0$ and $(\overline{x}, \overline{y}, \overline{p}^K)$ is a quasiequilibrium of $\mathcal{E}_{|K}$. Applying now Proposition 5.3.4, let for each K

$$\pi^K = \{(\pi_i^K)_{i \in I}, (\pi_j^K)_{j \in J}, \text{ and } \pi_0^K \text{ if } Z \neq \{0\}\}, \quad \overline{\pi}^K = \mathcal{R}_{\pi^K}$$

be such that each π_i^K, each π_j^K, if case π_0^K, and $\overline{\pi}^K$ belong to L', and

$$\pi_i^K \cdot \widehat{P}_i(\overline{x}) \geq \pi_i^K \cdot \overline{x}_i, \quad \pi_j^K \cdot \widehat{Y}_j(\overline{y}_j) \leq \pi_j^K \cdot \overline{y}_j, \quad i \in I,\ j \in J, \qquad (5.29)$$

$$\pi_0^K \cdot \widehat{Z}(\overline{z}) \leq \pi_0^K \cdot \overline{z} = 0 \text{ if } Z \neq \{0\}. \tag{5.30}$$

$$\forall i \in I, \forall x_i \leq \overline{x}_i,\ x_i \in Z_{\overline{x}_i} \cap K,\ \pi_i^K \cdot (\overline{x}_i - x_i) = \overline{\pi}^K \cdot (\overline{x}_i - x_i) = \overline{p}^K \cdot (\overline{x}_i - x_i), \tag{5.31}$$

$$\forall j \in J, \forall y_j \geq \overline{y}_j,\ y_j \in Z_{\overline{y}_j} \cap K,\ \pi_j^K \cdot (\overline{y}_j - y_j) = \overline{\pi}^K \cdot (\overline{y}_j - y_j) = \overline{p}^K \cdot (\overline{y}_j - y_j), \tag{5.32}$$

$$\forall z \geq \overline{z},\ z \in Z_{\overline{z}} \cap K,\ \pi_0^K \cdot (\overline{z} - z) = \overline{\pi}^K \cdot (\overline{z} - z) = \overline{p}^K \cdot (\overline{z} - z), \text{ if } Z \neq \{0\}, \tag{5.33}$$

$$\overline{\pi}^K \cdot (\omega - u) = \overline{p}^K \cdot (\omega - u) \text{ if } u \leq \omega, u \in \sum_{i \in I} Z_{\overline{x}_i} \cap K - \sum_{j \in J} Z_{\overline{y}_j} \cap K - Z_{\overline{z}} \cap K, \tag{5.34}$$

and finally, for every $i \in I$

$$\overline{\pi}^K \cdot \overline{x}_i = \overline{\pi}^K \cdot \omega_i + \sum_{j \in J} \theta_{i,j} \overline{\pi}^K \cdot \overline{y}_j. \tag{5.35}$$

From (5.29), (5.30) and properness, we deduce that for each i, j and, if case, 0, $\pi_i^K \cdot \omega \geq 0$, $\pi_j^K \cdot \omega \geq 0$, $\pi_0^K \cdot \omega \geq 0$. From (5.34) we deduce that $\overline{\pi}^K \cdot \omega = \overline{p}^K \cdot \omega > 0$. So at least one of π_i^K, π_j^K, π_0^K gives to ω a strictly positive value. We thus can normalize prices by the relation

$$(\sum_{i \in I} \pi_i^K + \sum_{j \in J} \pi_j^K + \pi_0^K) \cdot \omega = 1. \tag{5.36}$$

Since $\mathcal{E}$ is ω-proper at $(\overline{x}, \overline{y}, \overline{z})$, let V be a convex and circled τ-neighborhood of 0 such that each $\overline{x}_i + \omega + V \subset \widehat{P}_i(\overline{x})$, each $\overline{y}_j - (\omega + V) \subset \widehat{Y}_j$ and if $Z \neq \{0\}$, $\overline{z} - (\omega + V) \subset \widehat{P}_i(\overline{x})$.

From (5.29), (5.30) if $Z \neq \{0\}$, and (5.36), we deduce that $\pi_i^K \cdot V \leq \pi_i^K \cdot \omega \leq 1$ for each $i \in I$, $\pi_j^K \cdot V \leq \pi_j^K \cdot \omega \leq 1$ for each $j \in J$, and if $Z \neq \{0\}$, $\pi_0^K \cdot V \leq \pi_0^K \cdot \omega \leq 1$, which implies that all π_i^K, $i \in I$, π_j^K, $j \in J$, and if case π_0^K belong to V^0, the polar set of V in L'. Since L is locally convex, it follows from Alaoglu–Bourbaki's theorem that V^0 is τ-equicontinuous thus $\sigma(L', L)$-compact. Passing to subnets if necessary, we can assume that each $\pi_i^K \xrightarrow{\sigma(L',L)} \pi_i \in V^0$, each $\pi_j^K \xrightarrow{\sigma(L',L)} \pi_j \in V^0$, and if $Z \neq \{0\}$, $\pi_0^K \xrightarrow{\sigma(L',L)} \pi_0 \in V^0$. Denoting by π the collection of all π_i, π_j and if case π_0, let us define $\overline{\pi} = \mathcal{R}_\pi$ and observe that $\overline{\pi} \in L'$ follows from [SA].

By passing to limit in (5.36), we first get

$$(\sum_{i\in I} \pi_i + \sum_{j\in J} \pi_j + \pi_0) \cdot \omega = 1,$$

thus $\overline{\pi} \cdot \omega > 0$.

We now claim that $\overline{\pi}$ supports the allocation $(\overline{x}, \overline{y}, \overline{z})$. To see that, fix first $(x, y, z) \in \prod_{i\in I} P_i(\overline{x}) \times Y_j \times Z$. By construction of $\mathcal{K}$, all x_i, y_j, z belong to some $K_0 \in \mathcal{K}$ and thus to any $K \supset K_0$ in $\mathcal{K}$. It is the same for all $u_i = x_i \wedge \overline{x}_i$, $v_j = y_j \vee \overline{y}_j$ and $v_0 = z \vee \overline{z}$. Recalling that $Z_{\overline{x}_i}$, $Z_{\overline{y}_j}$, $Z_{\overline{z}}$ are lattices and using (5.29)–(5.33), we first get for each $K \in \mathcal{K}$, $K \supset K_0$,

$$\pi_i^K \cdot (x_i - u_i) \geq \pi_i^K \cdot (\overline{x}_i - u_i) = \overline{\pi}^K \cdot (\overline{x}_i - u_i),$$

$$\pi_j^K \cdot (y_j - v_j) \leq \pi_j^K \cdot (\overline{y}_j - v_j) = \overline{\pi}^K \cdot (\overline{y}_j - v_j),$$

and if $Z \neq \{0\}$,

$$\pi_0^K \cdot (z - v_0) \leq \pi_0^K \cdot (\overline{z} - v_0) = \overline{\pi}^K \cdot (\overline{z} - v_0).$$

Passing to limit in the three previous relations and using the definition of $\overline{\pi}$ and Lemma 5.3.2, we finally get

$$\overline{\pi} \cdot (x_i - u_i) \geq \pi_i \cdot (x_i - u_i) \geq \liminf_K (\overline{\pi}^K \cdot (\overline{x}_i - u_i)) \geq \overline{\pi} \cdot (\overline{x}_i - u_i),$$

$$\overline{\pi} \cdot (v_j - y_j) \geq \pi_j \cdot (v_j - y_j) \geq \liminf_K (\overline{\pi}^K \cdot (v_j - \overline{y}_j)) \geq \overline{\pi} \cdot (v_j - \overline{y}_j),$$

and if $Z \neq \{0\}$,

$$\overline{\pi} \cdot (v_0 - z) \geq \pi_0 \cdot (v_0 - z) \geq \liminf_K (\overline{\pi}^K \cdot (v_0 - \overline{z})) \geq \overline{\pi} \cdot (v_0 - \overline{z}),$$

which proves that $\overline{\pi} \cdot x_i \geq \overline{\pi} \cdot \overline{x}_i$ for each $i \in I$, $\overline{\pi} \cdot y_j \leq \overline{\pi} \cdot \overline{y}_j$ for each $j \in J$, and $\overline{\pi} \cdot z \leq \overline{\pi} \cdot \overline{z}$ if $Z \neq \{0\}$.

Letting (x, y, z) vary in $\prod_{i\in I} P_i(\overline{x}) \times Y_j \times Z$, we deduce that $\overline{\pi} \cdot P_i(\overline{x}) \geq \overline{\pi} \cdot \overline{x}_i$ for each $i \in I$, $\overline{\pi} \cdot Y_j \leq \overline{\pi} \cdot \overline{y}_j$ for each $j \in J$, and $\overline{\pi} \cdot Z \leq \overline{\pi} \cdot \overline{z}$ if $Z \neq \{0\}$, which proves the claim. Note that it follows from the fact that Z is a cone that $\overline{\pi} \cdot \overline{z} = 0$ and $\overline{\pi} \in Z^0$.

In order to prove that $(\overline{x}, \overline{y}, \overline{\pi})$ is a quasiequilibrium of $\mathcal{E}$, it remains to pass to limit in the relations (5.35). We now define for each i, $z_i = \overline{x}_i \wedge \omega_i$ and recall that each $Z_{\overline{x}_i}$ is a lattice containing $\overline{x}_i$ and ω_i thus that

$z_i \in Z_{\overline{x}_i}$. Since each $K \in \mathcal{K}$ is an order ideal containing all $\overline{x}_i$ and ω_i, we have $z_i \in Z_{\overline{x}_i} \cap K$ for each $i \in I$ and for each $K \in \mathcal{K}$. By a similar argument, $\overline{y}_j^+ \in Z_{\overline{y}_j} \cap K$ for each $j \in J$ and for each $K \in \mathcal{K}$.

For each $i \in I$, using (5.31) and (5.32), we can rewrite (5.35)

$$\pi_i^K \cdot (\overline{x}_i - z_i) + \sum_{j \in J} \theta_{i,j} \pi_j^K \cdot (\overline{y}_j^+ - \overline{y}_j) = \overline{\pi}^K \cdot (\omega_i - z_i) + \sum_{j \in J} \theta_{i,j} \overline{\pi}^K \cdot \overline{y}_j^+.$$

Passing to limit and using the definition of π and Lemma 5.3.2, we get for each $i \in I$:

$$\overline{\pi} \cdot (\overline{x}_i - z_i) + \sum_{j \in J} \theta_{i,j} \overline{\pi} \cdot (\overline{y}_j^+ - \overline{y}_j) \geq \liminf_K (\overline{\pi}^K \cdot (\omega_i - z_i) + \sum_{j \in J} \theta_{i,j} \overline{\pi}^K \cdot \overline{y}_j^+) \geq$$

$$\liminf_K \overline{\pi}^K \cdot (\omega_i - z_i) + \sum_{j \in J} \theta_{i,j} \liminf_K \overline{\pi}^K \cdot \overline{y}_j^+ \geq \overline{\pi} \cdot (\omega_i - z_i) + \sum_{j \in J} \theta_{i,j} \overline{\pi} \cdot \overline{y}_j^+.$$

That is, for each $i \in I$,

$$\overline{\pi} \cdot \overline{x}_i \geq \overline{\pi} \cdot \omega_i + \sum_{j \in J} \theta_{i,j} \overline{\pi} \cdot \overline{y}_j.$$

That all these inequalities are actually equalities comes from the fact that $\sum_{i \in J} \overline{x}_i - \sum_{j \in J} \overline{y}_j \in Z$ and that $\overline{\pi} \in Z^0$.

Finally, let us assume that for some $\lambda > 0$, $\lambda\omega \in \omega + \sum_{j \in J} Y_j + Z - \sum_{i \in I} X_i$. From $\overline{\pi} \cdot \omega > 0$ we deduce that there are some $x \in \prod_{i \in I} X_i$, some $y \in \prod_{j \in J} Y_j$, some $z \in Z$ such that

$$\overline{\pi} \cdot (\omega + \sum_{j \in J} y_j + z - \sum_{i \in I} x_i) > 0.$$

We then have for some $i_0 \in I$,

$$\overline{\pi} \cdot x_{i_0} < \overline{\pi} \cdot \omega_{i_0} + \sum_{j \in J} \theta_{i_0,j} \overline{\pi} \cdot y_j + \overline{\pi} \cdot z \leq \overline{\pi} \cdot \omega_{i_0} + \sum_{j \in J} \theta_{i_0,j} \overline{\pi} \cdot \overline{y}_j = \overline{\pi} \cdot \overline{x}_{i_0},$$

which proves that the quasiequilibrium $(\overline{x}, \overline{y}, \overline{\pi})$ is nontrivial. □

REMARK 5.3.5 The condition for nontriviality of the quasiequilibrium is in particular satisfied if $0 \in \sum_{j \in J} Y_j + Z - \sum_{i \in I} X_i$. Indeed from $\omega \in \sum_{i \in I} X_i - \sum_{j \in J} Y_j - Z$, that is, $-\omega \in \sum_{j \in J} Y_j + Z - \sum_{i \in I} X_i$, one deduces by convexity that for every $\lambda :\ 0 < \lambda < 1$, one has $(1-\lambda)(-\omega) \in \sum_{j \in J} Y_j + Z - \sum_{i \in I} X_i$, that is, $\lambda\omega \in \omega + \sum_{j \in J} Y_j + Z - \sum_{i \in I} X_i$.

The condition is consequently satisfied in exchange economies where consumption sets coincide with the positive cone. If inaction is possible as well for consumers as for producers, the condition is also satisfied in a production economy where consumption sets need not coincide with the positive cone.

Proposition 5.3.6 extends at several instances the equilibrium existence results of Mas-Colell–Richard [132] and Richard [154]: preferences are nonordered, consumption sets need not coincide with the positive cone of the commodity space; the quasiequilibrium existence does not require any kind of monotonicity or free-disposal assumption. For this last reason, Proposition 5.3.6 extends also Tourky [171] and [172].

Following an idea which goes to Podczeck [149], some of the previous techniques of proof allow to extend a result of Araujo–Monteiro [19] on the existence of equilibrium without uniform properness conditions.

5.3.5 The particular case of a strictly positive total initial endowment

We now assume that the commodity space (L, τ) has a quasi-interior point. More precisely, we consider an exchange economy

$$\mathcal{E} = (\langle L, L' \rangle, (L_+, P_i, \omega_i)_{i \in I})$$

without disposal activities, with the positive cone L_+ as consumption sets of the agents, and such that for each i, $\omega_i \in L_+$. We assume that $L(\omega)$, the principal order ideal generated by ω verifies $\operatorname{cl} L(\omega) = L$ (that is, $L(\omega)$ is τ-dense in L), that $\mathcal{E}$ has a nonempty fuzzy core and we start from $\overline{x} \in \mathcal{C}^f(\mathcal{E})$.

PROPOSITION 5.3.7 *If, in the previous conditions, $\mathcal{E}$ is pointwise F-proper at $\overline{x}$ with for each $i \in I$ a properness vector $v_i > 0$, and if each $P_i(\overline{x})$ is τ-open in L_+, then there exists $\overline{\pi} \in L'$ such that $(\overline{x}, \overline{\pi})$ is a nontrivial quasiequilibrium of $\mathcal{E}$.*

Proof. We first observe with Podczeck [149] that $\operatorname{cl}((L(\omega))_+) = L_+$. Indeed, if not, there is some $q \in L'$ and $z \in L_+ \setminus \operatorname{cl}((L(\omega))_+)$ such that $q \cdot z < 0 \leq q \cdot (L(\omega))_+$. Since, according to [SA] b, L' is a sublattice of $L^\sim$, $q^- = (-q) \vee 0$ and $q^+ = q \vee 0$ exist and belong to L'. Since $0 \leq q \cdot (L(\omega))_+$ and since $L(\omega)$ is an ideal of L, $q^- \cdot L(\omega) = 0$. From the density in L of $L(\omega)$, one deduces that $q^- \cdot L = 0$, that is, $q = q^+$. This obviously contradicts the existence of $z \in L_+$ such that $q \cdot z < 0$ and proves the observation.

Following this observation, there is no loss of generality to assume that for each $i \in I$, the properness vector v_i belongs to $(L(\omega))_+$. Set $v = \sum_{i \in I} v_i$, and let for each $i \in I$, U_i be an open convex properness neighborhood and Γ_i be the open convex cone generated by $v_i + \Gamma_i$.

Now a cautious reader can observe that with our hypothesis, under a slight modification of its proof, the conclusion of Proposition 5.3.5 can be replaced by: there exists $\overline{p} \in (L(\omega))^{\sim}$ such that $\overline{p} \cdot v > 0$ and $(\overline{x}, \overline{p})$ is a quasiequilibrium of $\mathcal{E}_{|_{L(\omega)}}$. As in Podczeck [149], it is easily seen that this quasiequilibrium is nontrivial.

The same cautious reader can observe that with our hypothesis and the same proof, the conclusion of Proposition 5.3.4 can be changed to: there exist for each i some $\pi_i \in L'$ and $\overline{\pi} = \bigvee_{i \in I} \pi_i$ such that $\pi_{|_{L(\omega)}} \leq \overline{p}$ and

$$\pi_i \cdot (\overline{x}_i + \Gamma_i) \geq \overline{\pi}_i \cdot \overline{x}_i = \overline{p} \cdot \overline{x}_i = \overline{p} \cdot \omega_i$$

holds for each $i \in I$.

From this, one easily deduces that $\overline{\pi}_{|_{L(\omega)}} = \overline{p}$, thus that $\overline{p}$ is τ continuous on $L(\omega)$. From the density in L of $L(\omega)$ one gets that $\overline{\pi}$ is the unique continuous extension of $\overline{p}$. From the density in L_+ of $(L(\omega))_+$ and from the τ-openness of $P_i(\overline{x})$ in L_+, one deduces that $(\overline{x}, \overline{\pi})$ is a quasiequilibrium of $\mathcal{E}$. Since the quasiequilibrium $(\overline{x}, \overline{p})$ of $\mathcal{E}_{|_{L(\omega)}}$ was nontrivial, the same is true of the quasiequilibrium $(\overline{x}, \overline{\pi})$ of $\mathcal{E}$. □

5.3.6 From quasiequilibrium to equilibrium

Adding Assumptions 1, 2, 3, 4 of Proposition 5.2.3, which insure the existence of consumption allocations in the fuzzy core of the economy, with either the interiority conditions (Proposition 5.3.1) or the different properness assumptions (Proposition 5.3.3 and its corollary, Propositions 5.3.6 and 5.3.7), which allow for their decentralization, yields several quasiequilibrium existence results. The conditions of Propositions 5.3.1 and 5.3.6 can likewise be added with the conditions of any nonemptiness result for the fuzzy core (see for example Remark 5.2.1). We leave the reader to summarize in each case the assumptions set on the commodity space and on the primitives of the economy. Let us only precise that interiority as well as pointwise F-properness or properness have to be assumed at each attainable allocation which could give an element of the fuzzy core, for example at every weakly Pareto optimal allocation.

Assume now that in either case the obtained quasiequilibrium $(\overline{x}, \overline{y}, \overline{\pi})$ is nontrivial (sufficient conditions have been stated in each case) and that Assumption 4 is satisfied. Noticing that Assumption 4 and the various irreducibility conditions stated in Chapter 2 do not depend on the vector space topology considered on the commodity space, we can repeat verbatim the arguments of Chapter 4 Section 4.3.3 and assert that, under any one of the irreducibility conditions proposed in Chapter 2, this nontrivial quasiequilibrium is actually an equilibrium.

5.4 A brief historical survey and a suggestion for a research agenda

The equilibrium existence study in infinite dimensional economies begins with a paper of Peleg and Yaari [148] devoted to allocation of resources over time, and unpublished notes of Debreu and Hildenbrand [62] on equilibrium under uncertainty.

Working in $\mathbb{R}^{\mathbb{N}}$, Peleg–Yaari's paper argues that, as linear functionals, prices should not be defined on the whole commodity space and that an extended equilibrium notion should be defined. Proving the existence of an allocation in the Debreu–Scarf core of the economy and decentralizing with prices this allocation, Peleg and Yaari prove the existence of an extended equilibrium in an exchange economy. A similar approach is accommodated to production economies by Boyd–McKenzie [46]. Extended equilibrium prices defined for more general settings are studied in several papers.

In contrast, the notes of Debreu and Hildenbrand claim that the equilibrium price should be a linear functional on the whole commodity space and conjecture equilibrium existence theorems under conditions analogue to the ones used in the finite dimensional setting. This chapter clearly belongs to this tradition and has attempted to give a systematic account of the results currently available.

The decisive steps in this line of research we had to report are:

1 Bewley's paper [35] and its equilibrium existence theorem in L_∞, extended and reproved several times between 1972 and 1986, extensively studied as to the economic implications of its different hypothesis;

2 Mas-Colell's paper [130], introducing the properness assumptions and allowing for equilibrium existence results in spaces L_p;

3 Mas-Colell–Richard's paper [132] which replaces the assumption of a topological vector lattice space by the today generally adopted assumptions [SA] on the commodity–price duality, allowing for an

equilibrium existence theorem in the space $M(K)$ equipped with the weak-star topology $\sigma(M(K), C(K))$.

Bewley's model and techniques are still a prototype for infinite horizon studies in models which depart from the standard general equilibrium model (for example, nonconvex economies or incomplete markets). Some of these models are also studied using properness assumptions. Our list of references is far from being representative in these respects.

In the framework of the standard model, Mas-Colell [130] was almost immediately extended in [181] to the case of an exchange economy with nonordered preferences and to economies with production in [153] and [182]. Mas-Colell–Richard's paper [132] was simultaneously extended to production economies in [154]. The remaining improvements concern the properness definition with Podczeck [149] and Tourky [171] and [172].

As to the proof strategy, Bewley [35] uses a limiting process on the equilibria of a net of finite dimensional economies. Under different forms, such limiting processes are a central ingredient of any proof in an infinite dimensional context.

Bewley [34], Magill [124], Mas-Colell [130], Mas-Colell–Richard [132], Richard [154], and many others use the Negishi [144] approach, an approach to the equilibrium existence problem as old as the first finite dimensional equilibrium existence theorems. Roughly speaking, the Negishi strategy of proof consists in establishing first an homeomorphism between the unit-simplex in the space of utilities (a finite dimensional space, since there are finitely many consumers) and the Pareto frontier of the set of attainable utility vectors. A point of the Pareto frontier corresponds to a weakly Pareto optimal consumption allocation which, in an infinite dimensional setting, can be price supported using uniform properness of preferences. The obtained quasiequilibrium relative to a price system determines in the space of utilities a vector of transfer-payments for each consumer. A (finite dimensional) fixed point argument shows that for some point of the unit-simplex the corresponding vector of transfer-payments is equal to 0, determining a quasiequilibrium. The advantage of this approach is to separate fixed point and separation arguments in the equilibrium existence proof. Its main drawback is to require that preferences be not only transitive and complete but monotone. They must also be represented by a continuous utility function. However in the finite dimensional case, the existence of a continuous utility function has no cost. In an infinite dimensional setting, as proved in [155], continuous, uniformly proper, complete preorders on L_+ can be represented by continuous utility functions.

This approach was used neither in this chapter nor in this book. Following [5], [171], [172], and as in [83], [64], [86], we have preferred to prove the equilibrium existence via the existence of Edgeworth equilibria. This strategy perfectly fulfills the objective of separating the fixed point argument and the separation argument, without the drawbacks of a Negishi proof. Using this approach and building on Tourky's properness definitions, we have reported the different results under as weak as possible assumptions.

Some work remains to do for clarifying the relations between uniform F-properness and Tourky's properness, in the direction of [14].

Moreover, in this chapter, the plausibility of the lattice structure of the commodity space was never questioned. The lattice structure was introduced in economic analysis by [3] and Mas-Colell [130] shows that it is in effect useful. However, this structure is missing in some plausible commodity spaces, and it is natural to look at what extent this hypothesis could be dropped when using properness assumptions. A first attempt was done in Aliprantis–Tourky–Yannelis [16] where the characterization of Pareto optimality and Edgeworth equilibrium is given using non-linear prices for decentralizing the relevant allocations. In [11] and [12], we go further and give necessary and sufficient conditions of compatibility between the topology and the order structure of a topological ordered space for a linear decentralization based on properness assumptions. Extension to production economies is still an open question.

Appendix A

In this appendix, we[1] summarize the most important continuity definitions and properties dealing with correspondences. In order to keep the notions as simple as possible, we will emphasize the particular case of metric spaces, though for some definitions, the general topological formulation may be more practical. We always assume that the spaces are Hausdorff.

In the Section A.1, we define the two notions of semicontinuity and give a characterization by sequences. Related concepts of continuity are introduced in Subsection A.2, and we establish the main relations with the concepts of the previous Sections. In Section A.3, we study different operations for correspondences (composition, closure, intersection, product, ...) which are very useful in Mathematical Economics. Finally, in the last section, we state the Maximum Theorem.

A.1 Upper and lower semicontinuity

Let us consider topological nonempty topological spaces X and Y. We first introduce convenient notations, that allow us to formulate an useful statement of the the notions of upper and lower semicontinuity.

DEFINITION A.1.1 *Let $\varphi : X \to Y$ be a correspondence, $B \subset Y$. We define:*
$\varphi^+(B) = \{x \in X \mid \varphi(x) \subset B\}$ (***superior inverse of*** φ),
$\varphi^-(B) = \{x \in X \mid \varphi(x) \cap B \neq \emptyset\}$ (***inferior inverse of*** φ).

PROPOSITION A.1.1 *Let $\varphi: X \to Y$ be a correspondence. The following properties are equivalent:*
i) φ is upper semicontinuous on X,
ii) For all open set $V \subset Y$, $\varphi^+(V)$ is open,
iii) For all closed $V \subset Y$, $\varphi^-(V)$ is closed.

Proof. Exercise left to the reader. □

[1]This appendix is written by Pascal Gourdel.

PROPOSITION A.1.2 *Let $\varphi: X \to Y$ be a correspondence. The following properties are equivalent:*
i) φ is lower semicontinuous on X,
ii) For all open set $V \subset Y$, $\varphi^-(V)$ is open,
iii) For all closed $V \subset Y$, $\varphi^+(V)$ is closed.

Proof. Exercise left to the reader. □

It is important to notice that the concept of lower (or upper) semicontinuity of a function do not coincide with the concept of lower (upper) semicontinuity of a correspondence. In the special case of a single-valued correspondence, it is possible to associate to this correspondence a function; in this special case, as it is stated in the next proposition, the two notions of semicontinuity for correspondences coincide and are equivalent to the classical continuity of the associated function.

PROPOSITION A.1.3 *Let $f: X \to Y$ be a function and let $\varphi : X \to Y$ the correspondence defined by: for all $x \in X$, $\varphi(x) = \{f(x)\}$. The following properties are equivalent:*
i) φ is lower semicontinuous on X,
ii) φ is upper semicontinuous on X,
iii) f is continuous on X.

Proof. Let us first remark that in this case, for any set $A \subset Y$ the inferior and superior inverse $\varphi^-(A)$, $\varphi^+(A)$, coincide with the classical notion $f^{-1}(A) = \{x \in X : f(x) \in A\}$. An easy application of Propositions A.1.1 and A.1.2, allows us to conclude. □

Since in a metric space the topology is entirely characterized by the notion of convergent sequences, it is possible to characterize the notions of semicontinuity by sequences. In the first result, we only assume that X is metric.

PROPOSITION A.1.4 *Let X be a metric space containing x. Let $\varphi : X \to Y$ be a correspondence.*
a) φ is upper semicontinuous at x if and only if for all convergent sequence $(x_k)_{k\in\mathbb{N}}$ convergent sequence to x, and for all V open neighborhood of $\varphi(x)$, there exists n such that for all $k \geq n$, $\varphi(x_k) \subset V$.
b) φ is lower semicontinuous at x if and only if for all $(x_k)_{k\in\mathbb{N}}$ convergent sequence to x, and V open set such that $\varphi(x) \cap V \neq \emptyset$, there exists n such that for all $k \geq n$, $\varphi(x_k) \cap V \neq \emptyset$.

Proof. a) ($\Rightarrow$) Let us suppose that $(x_k)_{k\in\mathbb{N}}$ converges to x.
Since φ is upper semicontinuous at x and since V is an open neighborhood of $\varphi(x)$, the set $\varphi^+(V)$ is a neighborhood of x. Recalling that x is the limit of a convergent sequence $(x_k)_{k\in\mathbb{N}}$, there exists n such that for all $k \geq n$, $x_k \in \varphi^+(V)$ or equivalently $\varphi(x_k) \subset V$.

a) ($\Leftarrow$) Assume the contrary, φ is not upper semicontinuous at x. There exists some neighborhood V of $\varphi(x)$ such that $\varphi^+(V)$ is not a neighborhood of x, hence one can construct $(x_k)_{k\in\mathbb{N}}$ convergent sequence to x, such that for all $k \in \mathbb{N}$, $x_k \notin \varphi^+(V)$ or equivalently $\varphi(x_k) \not\subset V$. A contradiction.

b) ($\Rightarrow$) Let us suppose that $(x_k)_{k\in\mathbb{N}}$ converges to x. Since φ is lower semicontinuous at x and since V is open and satisfies $\varphi(x) \cap V \neq \emptyset$, the set $\varphi^-(V)$ is a neighborhood of x. Recalling that x is the limit of a convergent sequence $(x_k)_{k\in\mathbb{N}}$, there exists n such that for all $k \geq n$, $x_k \in \varphi^-(V)$ or equivalently $\varphi(x_k) \cap V \neq \emptyset$.

b) ($\Leftarrow$) Assume the contrary, φ is not upper semicontinuous at x. There exists some open set V satisfying $\varphi(x) \cap V \neq \emptyset$ such that $\varphi^-(V)$ is not a neighborhood of x, hence one can construct $(x_k)_{k\in\mathbb{N}}$ convergent sequence to x, such that for all $k \in \mathbb{N}$, $x_k \notin \varphi^-(V)$ or equivalently $\varphi(x_k) \cap V = \emptyset$. A contradiction. □

In the case where the two spaces are metric spaces, we state the following theorem.

THEOREM A.1.1 : *Let $\varphi: X \to Y$ be a correspondence.*
a) If $\varphi(x)$ is compact, then φ is upper semicontinuous at x if and only if for all sequences $(x_k)_{k\in\mathbb{N}}$, $(y_k)_{k\in\mathbb{N}}$ such that $x_k \to x$, $y_k \in \varphi(x_k)$, there exists a subsequence $(y_{k_n})_{n\in\mathbb{N}}$ of $(y_k)_{k\in\mathbb{N}}$ and $y \in \varphi(x)$ such that $y_{k_n} \to y$.
b) φ is lower semicontinuous at x if and only if for all $(x_k)_{k\in\mathbb{N}}$ convergent sequence to x, and all $y \in \varphi(x)$, there exists a sequence $(y_k)_{k\geq k_0}$ in Y such that $y_k \to y$ and for all $k \geq k_0$, $y_k \in \varphi(x_k)$.

Proof. a) ($\Rightarrow$) Let us suppose that $(x_k)_{k\in\mathbb{N}}$ converges to x, $y_k \in \varphi(x_k)$, for all k, $(y_k)_{k\in\mathbb{N}}$ converges to $y \notin \varphi(x)$.

Let us first remark that for all $p \geq 1$, the set $V_p = \{y \in Y : d(y, \varphi(x)) < 1/p\}$ is an open neighborhood of $\varphi(x)$. In view of the previous proposition, there exists $n(p)$ such that for all $k \geq n(p)$, $\varphi(x_k)$ is a subset of V_p. We can assume with no loss of generality that the sequence $(n(p))_{p\in\mathbb{N}}$ is increasing, consequently for all $k \geq n(p)$, we can choose an integer $\psi(k)$ such that $n(\psi(k)) \leq k < n(\psi(k)+1)$. By construction, $\psi(k) \to +\infty$ when $k \to +\infty$.

Since for all $k \geq n(1)$, $y_k \in V_{\psi(k)}$, we can construct a sequence $(z_k)_{k\geq n(1)}$ of elements of $\varphi(x)$ such that $d(y_k, z_k) < 1/\psi(k)$. In view of the compactness of $\varphi(x)$, there exists a convergent subsequence $(z_{q_k})_k$ that converges to some element $y \in \varphi(x)$. By construction, the subsequence $(y_{q_k})_{k\geq n(1)}$ also converges to y.

($\Leftarrow$) Assume the contrary, that is, φ is not upper semicontinuous at point x. By definition, there exists some neighborhood V of $\varphi(x)$ such that for all open neighborhood U of x, there exists some x_U satisfying $\varphi(x_U) \not\subset V$. In particular for the base of neighborhood $U_k = B(x, 1/k)$, there exists some $x_k \in U_k$, $\varphi(x_k) \not\subset V$. This allows us to construct a sequence $(y_k)_{k\in\mathbb{N}}$ such that $y_k \in \varphi(x_k) \cap V^c$. Since V is open, its complementary V^c is closed and the sequence $(y_k)_{k\in\mathbb{N}}$ has no cluster point in V, in particular in $\varphi(x)$. A contradiction.

b) ($\Rightarrow$) Let us suppose that $(x_k)_{k\in\mathbb{N}}$ converges to x and that $y \in \varphi(x)$.

Let us first remark that for all $p \geq 1$, the set $V_p = B(y, 1/p)$ is an open set that intersects $\varphi(x)$. In view of the previous proposition, there exists $n(p)$ such that for all $k \geq n(p)$, $\varphi(x_k)$ intersects V_p. We can assume with no loss of generality that the sequence $(n(p))_{p\in\mathbb{N}}$ is increasing, consequently for all $k \geq n(p)$. we can choose an integer $\psi(k)$ such that $n(\psi(k)) \leq k < n(\psi(k)+1)$.

Since for all $k \geq n(1)$, $\varphi(x_k) \cap V_{\psi(k)} \neq \emptyset$, we can construct a sequence $(y_k)_{k\geq n(1)}$ of elements of $\varphi(x_k) \cap V_{\psi(k)}$. Since by construction, $\psi(k) \to +\infty$ when $k \to +\infty$, the sequence $(y_k)_{k\geq n(1)}$ converges to y.

($\Leftarrow$) Assume the contrary, that is, φ is not lower semicontinuous at point x. By definition, there exists some open set V containing a point $y \in \varphi(x)$ such that for all open neighborhood U of x there exists some x_U satisfying $\varphi(x_U) \cap V = \emptyset$. In particular for the base of neighborhood $U_k = B(x, 1/k)$, there exists some $x_k \in U_k$, $\varphi(x_k) \cap V = \emptyset$. Then for any sequence $(y_k)_{k\in\mathbb{N}}$ such that $y_k \in \varphi(x_k)$, one has $y_k \in V^c$. Since V is open, its complementary V^c is closed and the sequence $(y_k)_{k\in\mathbb{N}}$ has no cluster point in V, in particular y is not a cluster point. A contradiction. □

Either from Proposition A.1.1 or from Theorem A.1.1 (when the spaces are metric), it is possible to adapt for the correspondence case the Weierstrass theorem.

PROPOSITION A.1.5 *Let X be a compact space and $\varphi : X \to Y$ be a correspondence. If for all x in X, φ is upper semicontinuous at x and $\varphi(x)$ is compact, then the set $\varphi(X)$ is compact.*

Proof. Let $(V_i)_{i\in I}$ be an open covering of $\varphi(X)$. For all $x \in X$, since $\varphi(x)$ is a compact set, there exists a finite subcovering $(V_i)_{i\in I(x)}$. Defining the open set $O_x = \cup_{i\in I(x)} V_i$, it is immediate that $x \in \varphi^+(O_x)$. Moreover it follows from the upper semicontinuity of φ and from Proposition A.1.1, that the collection $(\varphi^+(O_x))_{x\in X}$ is an open covering of X. Since the set X is compact, we can consider a finite subcovering $(\varphi^+(O_x))_{x\in J}$ of X. It is easy to check that $\{V_i \ : \ i \in \bigcup_{x\in J} I(x)\}$ is a finite subcovering of $\varphi(X)$. This implies the compactness of $\varphi(X)$. □

A.2 Related concepts

DEFINITION A.2.1 *Let $\varphi: X \to Y$ be a correspondence. φ is said to be* ***open*** *(respectively* ***closed****) if $\mathcal{G}_\varphi$ is open (respectively closed) in $X \times Y$.*

DEFINITION A.2.2 *Let $\varphi: X \to Y$ be a correspondence. φ is said to have* ***open lower sections*** *(open lower preimage, open fibers), if*

$$\forall y \in Y, \quad \varphi^{-1}(y) \text{ is open in } X.$$

PROPOSITION A.2.1 *Let $\varphi: X \to Y$ be a correspondence. One has: φ open $\Longrightarrow$ φ has open lower sections $\Longrightarrow$ φ is lower semicontinuous.*

Proof. Exercise left to the reader. □

It is also possible to consider the concept of locally closed correspondence.

DEFINITION A.2.3 *Let $\varphi: X \to Y$ be a correspondence. φ is said* ***closed*** *at x if $y \notin \varphi(x)$ implies the existence of a neighborhood U_x, (respectively V_y) of x (respectively of y) such that $(U_x \times V_y) \cap \mathcal{G}_\varphi = \emptyset$.*

PROPOSITION A.2.2 *Let $\varphi: X \to Y$ be a correspondence.*
a) If φ is closed at x, $\varphi(x)$ is a closed subset of Y.
b) φ is closed in $X \times Y$ if and only if φ is closed at each point of X.

Proof. Exercise left to the reader. □

PROPOSITION A.2.3 *Let X, Y be metric spaces. Let $\varphi: X \to Y$ be a correspondence.*
φ is closed at x if and only if for all $(x_k)_{k\in\mathbb{N}}$ and $(y_k)_{k\in\mathbb{N}}$ such that $(x_k)_{k\in\mathbb{N}}$ converges to x, and $(y_k)_{k\in\mathbb{N}}$ converges to y, with $y_k \in \varphi(x_k)$, for all k, one has $y \in \varphi(x)$.

Proof. ($\Rightarrow$) Let us suppose that $(x_k)_k$ converges to x, $y_k \in \varphi(x_k)$, for all k, $(y_k)_{k\in\mathbb{N}}$ converges to $y \notin \varphi(x)$.

Since φ is closed at x, there exists a neighborhood U_x, (respectively V_y) of x (respectively of y) such that $(U_x \times V_y) \cap \mathcal{G}_\varphi = \emptyset$. For k, large enough, $(x_k, y_k) \in U_x \times V_y$, and consequently $y_k \notin \varphi(x_k)$. A contradiction.

($\Leftarrow$) Assume the contrary, that is φ is not closed at point x. By definition, there exists some some point $y \notin \varphi(x)$ such that for all open neighborhood U of x (respectively V of y), there exists some $(x_{U,V}, y_{U,V})$ in $(U \times V) \cap \mathcal{G}_\varphi$. In particular for the base of neighborhood $U_k = B(x, 1/k)$ (respectively $V_k = B(y, 1/k)$), there exists some $(x_k, y_k) \in U_k \times V_k$ such that $y_k \in \varphi(x_k)$. Since $x_k \to x$ and $y_k \to y$, at the limit, one gets $y \in \varphi(x)$. A contradiction. □

As shown by the following two propositions, the concepts of upper semicontinuity and closed correspondence are very close.

COROLLARY A.2.1 *Let X and Y be metric spaces. Let $\varphi: X \to Y$ be a correspondence.*
a) If φ is upper semicontinuous at $x \in X$ with $\varphi(x)$ closed, then φ is closed at x.
b) If φ is upper semicontinuous on X with closed values, then φ is closed in $X \times Y$.

Proof. Exercise left to the reader. □

A converse statement is more important in the applications.

PROPOSITION A.2.4 *Let $\varphi : X \to Y$ be a correspondence.*
a) If φ is closed at x and if there exists a neighborhood U_1 of x, and a compact set K such that $\varphi(U_1) \subset K$, then φ is upper semicontinuous at x.
b) If φ is closed and if there exists a compact set K such that $\varphi(X) \subset K$, then φ is upper semicontinuous on X.

Proof. a) Let V be an open neighborhood of $\varphi(x)$, we want to prove that the set $\varphi^+(V)$ is a neighborhood of x. By construction, the set $\tilde{K} = K \setminus \varphi(x)$ is compact.

Since φ is closed at point x, for all $y \in \tilde{K}$, there exists an open neighborhood U_y of x (respectively V_y of y) such that $\varphi(U_y) \subset (V_y)^c$. By a classical compactness argument, it is possible to consider a finite set $\{y_1, \dots, y_p\}$ such that the collection $(V_{y_i})_{i=1}^p$ is still a covering of $\tilde{K}$. If we denote by $\overline{U} = \cap_{i=1}^p U_{y_i}$, it is easy to check that $\overline{U}$ is a neighborhood of x and that $\tilde{K} \subset \overline{V}$.

We claim that $\overline{U} \subset \varphi^+(V)$, and in particular that $\varphi^+(V)$ is a neighborhood of x. Indeed, let us assume the contrary, there exists $x' \in \overline{U}$ and some $y \in \varphi(x') \cap V$. Since $\overline{U} \subset W$, $\varphi(x') \subset K$, $y \in \tilde{K}$, and consequently there exists some i such that $y \in V_{y_i}$. By definition of $\overline{U}$, it follows that $(x', y) \in (U_{y_i} \times V_{y_i}) \cap \mathcal{G}_\varphi$. A contradiction.
b) Obvious. □

A.3 Operations with correspondences

In this part, we will sometimes restrict ourselves to metric space thought it is possible to state Proposition A.3.2 and Corollary A.2.1 in a more general setting (topological space with convenient separation axioms), here these axioms are satisfied since the spaces are metric. Indeed we recall a classical topological result that is proved in [114], (this topological property is called "normal").

LEMMA A.3.1 *Let Z be a metric space and Z_1, Z_2 be closed subsets of Z. If $Z_1 \cap Z_2 = \emptyset$ then there exists some open neighborhood V_1 of Z_1 (respectively V_2 of Z_2) such that $V_1 \cap V_2 = \emptyset$.*

DEFINITION A.3.1 (***composition of correspondences***) *Let $\varphi: X \to Y$, and $\psi: Y \to Z$ be two correspondences. We define $\psi \circ \varphi$ by $(\psi \circ \varphi)(x) = \psi(\varphi(x))$.*

Roughly speaking, the property of semicontinuity is stable by composition. Formally one has:

PROPOSITION A.3.1 *Let $\varphi: X \to Y$ and $\psi: Y \to Z$ be two correspondences, $x \in X$,*
a) If φ is upper semicontinuous at x and ψ is upper semicontinuous on $\varphi(x)$, then $\psi \circ \varphi$ is upper semicontinuous at x.
b) If φ is lower semicontinuous at x and ψ is lower semicontinuous on $\varphi(x)$, then $\psi \circ \varphi$ is lower semicontinuous at x.

Proof. Exercise left to the reader. □

Under some condition, the property of semicontinuity is stable by the operation of closure.

DEFINITION A.3.2 (***closure of a correspondence***) *Let φ be a correspondence, its closure is the correspondence $\operatorname{cl}\varphi$ defined by: $\operatorname{cl}\varphi(x) = \operatorname{cl}(\varphi(x))$.*

PROPOSITION A.3.2 *Let $\varphi: X \to Y$ be a correspondence.*
a) If φ is lower semicontinuous at x, then $\operatorname{cl}\varphi$ is lower semicontinuous at x.
b) If Y is metric and if φ is upper semicontinuous at x, then $\operatorname{cl}\varphi$ is upper semicontinuous at x.

Proof. a) It suffices to remark that for any open set V of Y, for any x' in X, on has the equivalence between $V \cap (\operatorname{cl}\varphi)(x') \neq \emptyset$ and $V \cap \varphi(x') \neq \emptyset$. This can also be formulated as: the inverse inferior of φ and of $\operatorname{cl}\varphi$ coincide on open subsets.
b) Let V be an open neighborhood of $\operatorname{cl}\varphi(x)$, since $\operatorname{cl}\varphi(x) \cap V^c$ is empty, we can apply Lemma A.3.1 in order to get the existence of some open set V_1 and V_2 such that $\operatorname{cl}\varphi(x) \subset V_1 \subset V_2^c \subset V$. We claim that $\varphi^+(V_1) \subset (\operatorname{cl}\varphi)^+(V)$, and in particular that $(\operatorname{cl}\varphi)^+(V)$ is a neighborhood of x. Indeed, for all $x' \in \varphi^+(V_1)$, $\operatorname{cl}(\varphi)(x') \subset \operatorname{cl} V_1 \subset V_2^c \subset V$. □

DEFINITION A.3.3 (***intersection of correspondences***)*: Let φ and $\psi : X \to Y$ be two correspondences. We define $\varphi \cap \psi$ by $(\varphi \cap \psi)(x) = \varphi(x) \cap \psi(x)$.*

PROPOSITION A.3.3 *Let φ and $\psi: X \to Y$ be two correspondences, $x \in X$*
a) If φ is closed at x, ψ is upper semicontinuous at x, and $\psi(x)$ is compact, then $\varphi \cap \psi$ is upper semicontinuous at x and the set $(\varphi \cap \psi)(x)$ is compact.
b) If φ is lower semicontinuous at x and ψ is open, then $\varphi \cap \psi$ is lower semicontinuous at x.
c) Let Y be a metric space. If φ and ψ are upper semicontinuous at x and if $\varphi(x)$ and $\psi(x)$ are closed, then $\varphi \cap \psi$ is upper semicontinuous at x.

Proof. a) Let W be an open neighborhood of $\varphi(x) \cap \psi(x)$. For all $y \in \psi(x) \setminus \varphi(x)$, since φ is closed at x, there exists a neighborhood U_y of x (respectively V_y of y) such that $(U_y \times V_y) \cap \mathcal{G}_\varphi = \emptyset$. By a classical compactness argument, it is possible to consider a finite set $\{y_1, \ldots, y_p\}$ such that the collection $(V_{y_i})_{i=1}^p$ is still a covering of $\varphi(x) \setminus W$. If we denote by $\overline{U} = \cap_{i=1}^p U_{y_i}$ and by $V = \cup_{i=1}^p V_{y_i}$, it is easy to check that $\overline{U}$ is a neighborhood of x and that $\varphi(x) \subset W \cup \overline{V}$. Consequently, using the upper semicontinuity of ψ, the set $\tilde{U} = \psi^+(W \cup V)$ is a neighborhood of x.

We claim that $(\overline{U} \cap \tilde{U}) \subset (\varphi \cap \psi)^+(W)$, and in particular that $(\varphi \cap \psi)^+(W)$ is a neighborhood of x. Indeed, let us assume the contrary, there exists $x' \in \overline{U} \cap \tilde{U}$ and some $y \in (\varphi \cap \psi)(x') \cap W^c$. Since $x' \in \tilde{U}$, $\varphi(x') \subset W \cup V$, consequently $y \in V$. There exists some i such that $y \in V_{y_i}$. By definition of $\overline{U}$, it follows that $(x', y) \in (U_{y_i} \times V_{y_i}) \cap \mathcal{G}_\varphi$. A contradiction.

b) Let W be an open set satisfying $\varphi(x) \cap \psi(x) \neq \emptyset$. For ψ is open, there exists a neighborhood U_x of x (respectively V_y of y) such that $(U_x \times V_y) \subset \mathcal{G}_\psi$. Since φ is upper semicontinuous, the set $\varphi^-(W)$ is an open neighborhood of x. It is easy to check that $(U_x \cap \varphi^-(W)) \subset (\varphi \cap \psi)^-(W)$ and the conclusion holds.

c) Let W be an open neighborhood of $\varphi(x) \cap \psi(x)$. Since $\varphi(x) \setminus W$ and $\varphi(x) \setminus W$ are closed with an empty intersection, we can apply Lemma A.3.1 in order to get the existence of some open neighborhood V_1 of $\varphi(x) \setminus W$ (respectively V_2 of $\psi(x) \setminus W$).

We claim that $\varphi^+(W \cup V_1) \cap \psi^+(W \cup V_2) \subset (\varphi \cap \psi)^+(W)$, and in particular that $(\varphi \cap \psi)^+(W)$ is a neighborhood of x. Indeed, for all $x' \in \varphi^+(W \cup V_1) \cap \psi^+(W \cup V_2)$, and some $(\varphi \cap \psi)(x') \subset ((W \cup V_1) \cap \psi^+(W \cup V_2)) = W$. □

DEFINITION A.3.4 (***product of correspondences***)*: Let $(\varphi_i)_{i=1}^n : X \to Y_i$, be n correspondences. We define $\prod_{i=1}^n \varphi_i : X \to \prod_{i=1}^n Y_i$ by $(\prod_{i=1}^n \varphi_i)(x) = \prod_{i=1}^n \varphi_i(x)$ for all $x \in X$.*

PROPOSITION A.3.4 *Let $(\varphi_i)_{i=1}^n: X \to \prod_{i=1}^n Y_i$ be n correspondences, $x \in X$.*
a) If for all i, φ_i is lower semicontinuous at x, then $\prod_{i=1}^n \varphi_i$ is lower semicontinuous at x,
b) If for all i, φ_i is upper semicontinuous at x and $\varphi_i(x)$ is compact, $\prod_{i=1}^n \varphi_i$ is upper semicontinuous at x.

Proof. We first prove the results when $n = 2$. By an easy induction argument, we can deduce the general case.
a) Let V be an open subset of $Y_1 \times Y_2$ such that there exists $y \in (\varphi_1(x) \times \varphi_2(x)) \cap V$. If we denote by (y_1, y_2) the components of y, by definition of the product topology, there exists an open neighborhood V_1 of y_1 (respectively V_2 of y_2) such that $(V_1 \times V_2) \subset V$. It is easy to check that $(\varphi_1^-(V_1) \cap \varphi_2^-(V_2)) \subset (\varphi_1 \times \varphi_2)^-(V)$ in particular that $(\varphi_1 \times \varphi_2)^-(V)$ is a neighborhood of x.

b) Though the result is true for any topological space, we only give here the proof for the metric spaces. Let $(x_k)_{k\in\mathbb{N}}$, $(y_k)_{k\in\mathbb{N}}$ such that $x_k \to x$, $y_k^1 \in \varphi_1(x_k)$, $y_k^2 \in \varphi_2(x_k)$. Since φ_1 is upper semicontinuous, there exists a subsequence $(y_{k_n}^1)_{n\in\mathbb{N}}$ of $(y_k^1)_{k\in\mathbb{N}}$ and $y^1 \in \varphi_1(x)$ such that $y_{k_n}^1 \to y^1$. Since φ_2 is upper semicontinuous, there exists a subsequence $(y_{p_n}^2)_{n\in\mathbb{N}}$ of $(y_{k_n}^2)_{n\in\mathbb{N}}$ and $y^2 \in \varphi_2(x)$ such that $y_{k_n}^2 \to y^2$. By construction, the subsequence $((y_{p_n}^1, y_{p_n}^2))_{n\in\mathbb{N}}$ of $(y_k)_{k\in\mathbb{N}}$ converges to $(y^1, y^2) \in (\varphi_1 \times \varphi_2)(x)$, and the conclusion holds in view of Theorem A.1.1. □

DEFINITION A.3.5 (***Sum and convex hull of correspondences***) *Let $(\varphi_i)_{i=1}^n$ be n correspondences, $x \in X$. We define:*
the sum of $(\varphi_i)_{i=1,\dots,n}$: $(\sum_{i=1}^n \varphi_i)(x) = \sum_{i=1}^n \varphi_i(x)$
the convex hull of φ: $\operatorname{co}\varphi(x) = \operatorname{co}(\varphi(x))$.

PROPOSITION A.3.5 : *Let $(\varphi_i)_{i=1}^n : X \to Y$, be n correspondences, $x \in X$.*
a) If $(\varphi_i)_{i=1,\dots,n}$ are lower semicontinuous at x, their sum is lower semicontinuous at x.
b) If $(\varphi_i)_{i=1,\dots,n}$ are upper semicontinuous at x, and for all i, $\varphi_i(x)$ is compact, then their sum is upper semicontinuous at x.

Proof. We first prove the results when $n = 2$. By an easy induction argument, we can deduce the general case. Let us define $\psi : Y \times Y \to Y$, such that $\psi(y_1, y_2) = y_1 + y_2$, ψ is obviously continuous. Remark that $\varphi_1 + \varphi_2 = \psi \circ (\varphi_1 \times \varphi_2)$.
a) In view of this remark, $\varphi_1 + \varphi_2$ is lower semicontinuous since the product of lower semicontinuous correspondences is lower semicontinuous (Proposition A.3.4) and since this property is stable by composition (Proposition A.3.1).
b) In view of this remark, $\varphi_1 + \varphi_2$ is upper semicontinuous since here the product $\varphi_1 \times \varphi_2$ is upper semicontinuous (Proposition A.3.4) and since this property is stable by composition (Proposition A.3.1). □

PROPOSITION A.3.6 *Let Y be a convex subset of $\mathbb{R}^\ell$ and $\varphi : X \to Y$, be a correspondence, and $x \in X$,*
a) If φ is lower semicontinuous at x, $\operatorname{co}\varphi$ is lower semicontinuous at x.
b) If φ is upper semicontinuous at x and $\varphi(x)$ is compact, then $\operatorname{co}\varphi$ is upper semicontinuous at x.

Proof. Let V be an open subset of Y such that there exists $y \in (\operatorname{co}\varphi(x)) \cap V$. There exists some n, $y_1, \dots, y_n$ and $\alpha_1, \dots, \alpha_n$ such that $y = \alpha_1 y_1 + \dots + \alpha_n y_n$, $y_i \in \varphi(x)$, $\alpha_i \in [0, 1]$ and $\alpha_1 + \dots + \alpha_n = 1$. If we define $\Gamma : Y^n \to Y$ by $\Gamma(z_1, \dots, z_n) = \alpha_1 z_1 + \dots + \alpha_n z_n$, since Γ is upper semicontinuous, for all i, there exists V_i open neighborhood of y_i such that $\Gamma(V_1 \times \dots \times V_n) \subset V$. It is easy to check that $\cap_{i=1}^n \varphi^-(V_i) \subset (\operatorname{co}\varphi)^-(V)$ in particular that $(\operatorname{co}\varphi)^-(V)$ is a neighborhood of x.
b) Let us first remark that it follows from Caratheodory's Theorem that $(\operatorname{co}\varphi)(x)$ is compact. Let V be an open neighborhood of $\operatorname{co}\varphi(x)$. There exists $\tilde{V}$ open and convex such that $\operatorname{co}\varphi(x) \subset \tilde{V} \subset V$. Since φ is upper semicontinuous, the set $\varphi^+(\tilde{V})$ is an open neighborhood of x. The conclusion is a consequence of the inclusions: for all $x' \in \varphi^+(\tilde{V})$, $\operatorname{co}\varphi(x') \subset \operatorname{co}\tilde{V} \subset V$. □

A.4 Maximum Theorem

In Mathematical Economics, one has frequently to deal with optimization problems and more precisely one has to take care about the topological dependence of both the set of optimal solutions and the optimal value of the problem when the domain of optimization varies.

Let us consider the problem of optimization:

$$P_x : g(x) = \max \{f(x,y) \mid y \in \varphi(x)\}$$

where $f : X \times Y \to \mathbb{R}$ is function and $\varphi : X \to Y$ is a correspondence. The real number $g(x)$ is the optimal value of P_x and the correspondence of optimal solutions $\psi : X \to Y$, is defined by: $\psi(x) = \{y \in \varphi(x) \mid f(x,y) = g(x)\}$.

THEOREM A.4.1 (BERGE) *With the previous notations, if f is a continuous function and φ is lower semicontinuous and upper semicontinuous, with nonempty compact values, then*
a) the correspondence of optimal solutions ψ is upper semicontinuous, with compact values.
b) the value function g is continuous.

Proof. a) Let $\lambda(x) = \{y \in Y \ : \ f(x,y) \geq f(x,z)$, for all $z \in \varphi(x)\}$. Note that $\psi = \varphi \cap \lambda$, hence in order to prove the upper semi-continuity, it is enough to prove that λ is closed and to use Proposition A.3.3 recalling that φ is upper semicontinuous with nonempty compact values.

Let us now prove that λ is closed, indeed if $y \notin \lambda(x)$, there exist $m \in \mathbb{R}$ and $z \in \varphi(x)$ such that $f(x,y) < m < g(x) = f(x,z)$. The continuity of f allow us to prove the existence of some neighborhood U_x of x (respectively V_y of y, V_z of z, such that

$$\text{for all } x' \in U, \text{ for all } y' \in V_y, \text{ for all } z' \in V_z, \text{ one has } f(x',y') < m < f(x',z').$$

In view of the lower semicontinuity of φ, there exists U such that for all $x' \in U$, there exists $z' \in V_z \cap \varphi(x')$. Consequently, for $(x',y') \in (U_x \cap U) \times V_y$, there exists $z' \in V_z \cap \varphi(x')$ such that $f(x',y') < m < f(x',z') \leq g(x')$. It follows that $((U_x \cap U) \times V_y) \cap \mathcal{G}_\lambda$ is empty, and the conclusion holds.

b) Let us first remark that if we denote by I the identity of Y, I is upper semicontinuous with nonempty compact values. It is easy to check that $g = f \circ (I \times \psi)$. In view of Propositions A.3.4 and A.3.1, recalling that we already proved that ψ is upper semicontinuous with compact values, one deduces that g is upper semicontinuous. Since g is function, the continuity of g is proved (cf. Proposition A.1.3).

References

[1] C. D. Aliprantis. On the Mas-Colell–Richard equilibrium theorem. *J. Econom. Theory*, 74:414–424, 1997.

[2] C. D. Aliprantis and K. C. Border. *Infinite Dimension Analysis: A Hitchhiker's Guide.* Springer–Verlag, Heidelberg and New York, 1999. 2nd. edition.

[3] C. D. Aliprantis and D. J. Brown. Equilibria in markets with a Riesz space of commodities. *J. Math. Econom.*, 11:189–207, 1983.

[4] C. D. Aliprantis, D. J. Brown, and O. Burkinshaw. An economy with infinite dimensional commodity space. *Econom. Lett.*, 23:1–4, 1987.

[5] C. D. Aliprantis, D. J. Brown, and O. Burkinshaw. Edgeworth equilibria. *Econometrica*, 55:1109–1137, 1987.

[6] C. D. Aliprantis, D. J. Brown, and O. Burkinshaw. Edgeworth equilibria in production economies. *J. Econom. Theory*, 43:252–291, 1987.

[7] C. D. Aliprantis, D. J. Brown, and O. Burkinshaw. *Existence and Optimality of Competitive Equilibria.* Springer–Verlag, Heidelberg and New York, 1990.

[8] C. D. Aliprantis and O. Burkinshaw. *Locally Solid Riesz Spaces.* Academic Press, New York, 1978.

[9] C. D. Aliprantis and O. Burkinshaw. The fundamental theorems of welfare economics without proper preferences. *J. Math. Econom.*, 17:41–54, 1988.

[10] C. D. Aliprantis and O. Burkinshaw. When is the core equivalence theorem valid. *Econom. Theory*, 1:169–182, 1991.

[11] C. D. Aliprantis, M. Florenzano, and R. Tourky. General equilibrium analysis in ordered topological vector spaces. Technical Report 128, CERMSEM, Université de Paris 1, Paris, 2002.

[12] C. D. Aliprantis, M. Florenzano, and R. Tourky. Linear and non-linear price decentralization. Technical report, CERMSEM, Université de Paris 1, Paris, 2003.

[13] C. D. Aliprantis, P. K Monteiro, and R. Tourky. Non-marketed options, non-existence of equilibria, and non-linear prices. *J. Econom. Theory*, forthcoming.

[14] C. D. Aliprantis, R. Tourky, and N. C. Yannelis. Cone conditions in general equilibrium theory. *J. Econom. Theory*, 92:96–121, 2000.

[15] C. D. Aliprantis, R. Tourky, and N.C. Yannelis. The Riesz–Kantorovich formula and general equilibrium theory. *J. Math. Econom.*, 34:55–76, 2000.

[16] C. D. Aliprantis, R. Tourky, and N.C. Yannelis. A theory of value with non-linear prices: equilibrium analysis beyond vector lattices. *J. Econom. Theory*, 100:22–72, 2001.

[17] N. Allouch and M. Florenzano. Edgeworth and Walras equilibria of an arbitrage-free economy. *Econom. Theory*, forthcoming.

[18] A. Araujo. Lack of pareto optimality in economies with infinitely many commodities: The need for impatience. *Econometrica*, 53:455–461, 1985.

[19] A. Araujo and P. K. Monteiro. Equilibrium without uniform conditions. *J. Econ. Theory*, 48:416–427, 1989.

[20] K. J. Arrow. An extension of the basic theorems of the classical welfare economics. In *Proceedings of the Second Berkeley Symposium on Mathematical Statistics and Probability*, pages 507–532. University of California Press, 1951.

[21] K. J. Arrow and G. Debreu. Existence of equilibrium for a competitive economy. *Econometrica*, 22:265–290, 1954.

[22] K. J. Arrow and F. H. Hahn. *General Competitive Analysis.* Holden-Day, San Francisco, 1971.

[23] J. P. Aubin. *Mathematical Methods of Game and Econom. Theory.* North–Holland, Amsterdam and New York, 1979.

[24] K. Back. Structure of consumption sets and existence of equilibrium in infinite-dimensional spaces. *J. Math. Econ.*, 17:89–99, 1988.

[25] H. Ben-El-Mechaiek, P. Deguire, and A. Granas. Points fixes et coïncidences pour les applications multivoques (applications de ky fan). *C.R. Acad. Sc. Paris*, Série I Math. 295, 1982.

[26] H. Ben-El-Mechaiek, P. Deguire, and A. Granas. Points fixes et coïncidences pour les applications multivoques II (applications de type φ and φ^*). *C.R. Acad. Sc. Paris*, Série I Math. 295, 1982.

[27] C. Berge. *Espaces topologiques et fonctions multivoques.* Dunod, Paris, 1959.

[28] G. M. Bergman and B. R. Halpern. A fixed point theorem for inward and outward maps. *Trans. Amer. Math. Soc.*, 130:353–358, 1968.

[29] T. C. Bergstrom. The existence of maximal elements and equilibria in the absence of transitivity. Technical report, University of Michigan, 1975.

[30] T. C. Bergstrom. How to discard "free disposability" - at no cost. *J. Math. Econom.*, 3:131–134, 1976.

[31] T. C. Bergstrom, R. P Parks, and T. Rader. Preferences which have open graphs. *J. Math. Econom.*, 3:265–268, 1976.

[32] M. Berliant and T. ten Raa. A foundation of location theory: consumer preferences and demand. *J. Econom. Theory*, 44:336–353, 1988.

[33] M. Besada, M. Estevez, and C. Herves. Equilibria in economies with infinitely many commodities. *Econom. Lett.*, 26:203–207, 1988.

[34] T. F. Bewley. A theorem on the existence of competitive equilibria in a market with a finite number of agents and whose commodity space is L_∞. Technical report, Core Discussion Paper, Université de Louvain, 1969.

[35] T. F. Bewley. Existence of equilibria in economies with infinitely many commodities. *J. Econom. Theory*, 4:514–540, 1972.

[36] V. Boehm. The core of an economy with production. *Rev. Econom. Stud.*, 41:429–436, 1974.

[37] H. F. Bohenblust and S. Karlin. On a theorem of Ville. In H. W. Kuhn and A. W. Tucker, editors, *Contributions to the theory of games, Vol. I*, Ann. of Math. Stud. 24, pages 155–160. Princeton University Press, 1950.

[38] P. Bojan. A generalization of theorems on the existence of competitive economic equilibria to the case of infinitely many commodities. *Math. Balkanica*, 4:490–494, 1974.

[39] K. C. Border. A core existence theorem for games without ordered preferences. *Econometrica*, 52:1537–1542, 1984.

[40] K.C. Border. *Fixed point theorems with applications to economics and game theory.* Cambridge U. Press, 1985.

[41] A. Borglin and H. Keiding. Existence of equilibrium actions and of equilibrium. *J. Math. Econom.*, 3:313–316, 1976.

[42] N. Bourbaki. *Elements de Mathématiques, Livre V. Espaces vectoriels topologiques, Chapitres 1,2.* Hermann, 1966. Augmented and revised edition. English translation published by Springer–Verlag.

[43] N. Bourbaki. *Elements de Mathématiques. Topologie générale I, Chapitres 1 à 4.* Hermann, Paris, 1971. First edition: 1940–1949. English translation published by Reading, Massachussets: Addison–Wesley.

[44] N. Bourbaki. *Elements de Mathématiques, Livre V. Espaces vectoriels topologiques, Chapitres 3, 4, 5.* Hermann, Paris, 1973. First edition: 1964. English translation published by Springer–Verlag.

[45] N. Bourbaki. *Elements de Mathématiques. Topologie générale II, Chapitres 5 à 10.* Hermann, Paris, 1974. First edition: 1940–1949. English translation published by Reading, Massachussets: Addison–Wesley.

[46] J. H. Boyd and L. W. McKenzie. The existence of competitive equilibrium over in infinite horizon with production and general consumption sets. *International Economic Review*, 34:1–20, 1993.

[47] L. E. J. Brouwer. On continuous one-to-one transformations of surfaces into themselves. *Proc. Kon. Ned. Ak. V. Wet. Ser. A*, 11:788–798, 1909.

[48] L. E. J. Brouwer. On continuous one-to-one transformations of surfaces into themselves. *Proc. Kon. Ned. Ak. V. Wet. Ser. A*, 14:300–310, 1912.

[49] F. Browder. The fixed point theory of multivalued mappings in topological vector spaces. *Math. Ann.*, 177:283–301, 1968.

[50] D. J. Brown and L. Lewis. Myopic economic agents. *Econometrica*, 49:359–368, 1981.

[51] C. Castaing and M. Valadier. *Convex Analysis and Measurable Multifunctions.* Lecture Notes in Math. Springer–Verlag, Heidelberg and New York, 1977.

[52] A. Cellina. Approximation of set valued functions and fixed point theorems. *Ann. Mat. Pura Appl.*, IV, LXXXII:17–24, 1969.

[53] A. Cellina. A theorem on the approximation of compact multivalued mappings. *Atti. Acad. Naz. Lincei*, 8:149–153, 1969.

[54] B. Cornet. Fixed points and surjectivity theorems for correspondences; applications. Technical report, Université de Paris-Dauphine, Paris, 1975.

[55] G. Debreu. A social equilibrium existence theorem. *Proc. Natl. Acad. Sci. USA 38*, pages 886–893, 1952.

[56] G. Debreu. Valuation equilibrium and Pareto optimum. *Proc. Natl. Acad. Sci. USA 40*, pages 588–592, 1954.

[57] G. Debreu. Market equilibrium. *Proc. Natl. Acad. Sci. USA 42*, pages 876–878, 1955.

[58] G. Debreu. *Theory of Value - An Axiomatic Analysis of Economic Equilibrium.* Cowles Foundation Monograph 17. Wiley, New York, 1959.

[59] G. Debreu. New concepts and techniques for equilibrium analysis. *International Economic Review*, 3:257–272, 1962.

[60] G. Debreu. Continuity properties of Paretian utility. *International Economic Review*, 5:285–293, 1964.

[61] G. Debreu. Four aspects of the mathematical theory of economic equilibrium. *Proceedings of the International Congress of Mathematicians, Vancouver*, pages 65–77, 1974.

[62] G. Debreu and W. Hildenbrand. Unpublished notes on the existence of equilibria in markets with infinite dimensional commodity spaces. Technical report, University of California, Berkeley, 1968. Unpublished notes.

[63] G. Debreu and H. Scarf. A limit theorem on the core of an economy. *International Economic Review*, 4:235–246, 1963.

[64] M. Deghdak and M. Florenzano. Decentralizing Edgeworth equilibria in economies with many commodities. *Econom. Theory*, 14:297–310, 1999.

[65] D. Duffie. Competitive equilibria in general choice spaces. *J. Math Econom.*, 14:1–23, 1986.

[66] D. Duffie. *Dynamic Asset Pricing Theory.* Princeton University Press, Princeton and Oxford, 2001. 3rd. edition.

[67] J. Dugundji and A. Granas. *Fixed Point Theory, Vol. 1.* Polish Scientific Publishers, Warsaw, 1982.

[68] N. Dunford and J. T. Schwartz. *Linear Operators. Part 1: General Theory.* Interscience Publishers, New York, 1966. 3rd. edition.

[69] F. Y. Edgeworth. *Mathematical Psychics.* Kegan Paul, London, 1881.

[70] M. Esteban-Bravo and P. Gourdel. On general equilibrium in economies with externalities. Technical Report 136, CERMSEM, Université de Paris 1, Paris, 2002.

[71] K. Fan. Fixed points and minimax theorems in locally convex topological linear spaces. *Proc. Natl. Acad. Sci. USA 38*, pages 121–126, 1952.

[72] K. Fan. A generalization of Tichonoff's fixed point theorem. *Math. Ann.*, 142:305–310, 1961.

[73] K. Fan. Extension of two fixed point theorems of F. E. Browder. *Math. Z.*, 112:234–240, 1969.

[74] K. Fan. Extension of two fixed point theorems of Browder. In A. Dold and B. Eckman, editors, *Set-valued mappings, selections and topological properties of* 2^X, Lecture Notes in Math. 171, pages 12–16. Springer–Verlag, Berlin Heidelberg New York, 1970.

[75] K. Fan. A minimax inequality and applications. In O. Shisha, editor, *Inequalities III*, pages 103–113. Academic Press, New York and London, 1972.

[76] P. C. Fishburn. *Utility Theory for Decision Making.* Wiley, New York, 1970.

[77] M. Florenzano. Quasi-equilibrium in abstract economies without ordered preferences. Technical Report 8019, CEPREMAP, Paris, 1980.

[78] M. Florenzano. *L'Équilibre Économique Général Transitif et Intransitif : Problèmes d'Éxistence.* Monographies du Séminaire d'Économétrie XVI. Éditions du CNRS, Paris, 1981.

[79] M. Florenzano. The Gale–Nikaido–Debreu lemma and the existence of transitive equilibrium with or without the free-disposal assumption. *J. Math. Econom.*, 9:113–134, 1982.

[80] M. Florenzano. On the existence of equilibria in economies with an infinite dimensional commodity space. *J. Math. Econom.*, 12:207–219, 1983.

[81] M. Florenzano. Equilibrium in a production economy on an infinite dimensional commodity space. A unifying approach. Technical Report 8740, CEPREMAP, Paris, 1987.

[82] M. Florenzano. On the non-emptiness of the core of a coalitional production economy without ordered preferences. *J. Math. Anal. Appl.*, 141:484–490, 1989.

[83] M. Florenzano. Edgeworth equilibria, fuzzy core and equilibria of a production economy without ordered preferences. *J. Math. Anal. Appl.*, 153:18–36, 1990.

[84] M. Florenzano. Quasiequilibria in abstract economies: Application to the overlapping generations model. *J. Math. Anal. Appl.*, 182:616–636, 1994.

[85] M. Florenzano, P. Gourdel, and A. Jofre. Supporting weakly pareto optimal allocations in infinite dimensional nonconvex economies. Technical Report 2002.129, CERMSEM, Université de Paris 1, Paris, 2002.

[86] M. Florenzano and V. M. Marakulin. Production equilibria in vector lattices. *Econom. Theory*, 17:577–598, 2001.

[87] M. Florenzano and C. Le Van. *Finite Dimensional Convexity and Optimization.* Studies in Econom. Theory 13. Springer–Verlag, Heidelberg and New York, 2001.

[88] M. Florig. Contributions à la théorie de l'équilibre général. Technical report, Université de Paris 1, Paris, 1997. Ph.D thesis.

[89] M. Florig. On irreducible economies. *Annales d'Économie et de Statistiques*, 61:183–199, 2001.

[90] V. Fon and Y. Otani. Classical welfare theorems with non transitive and non complete preferences. *J. Econom. Theory*, 20:409–418, 1979.

[91] D. Gale. The law of supply and demand. *Math. Scand.*, 3:155–169, 1955.

[92] D. Gale. Price equilibrium for linear models of exchange. Technical Report P-1156, The Rand Corporation, 1957.

[93] D. Gale. The linear exchange model. *J. Math. Econom.*, 3:205–209, 1976.

[94] D. Gale and A. Mas-Colell. An equilibrium existence theorem for a general model without ordered preferences. *J. Math. Econom.*, 2:9–15, 1975.

[95] D. Gale and A. Mas-Colell. On the role of complete transitive preferences in equilibrium theory. In G. Schwödiauer, editor, *Equilibrium and Disequilibrium in Econom. Theory*, pages 7–14. D. Reidel Publishing Company, Dordrecht Holland, 1977.

[96] D. Gale and A. Mas-Colell. Corrections to an equilibrium existence theorem for a general model without ordered preferences. *J. Math. Econom.*, 6:297–298, 1979.

[97] I. L. Glicksberg. A further generalization of the Kakutani fixed point theorem with applications to Nash equilibrium points. *Proc. Amer. Math. Soc. 3*, pages 170–174, 1952.

[98] P. Gourdel. Existence of intransitive equilibria in non-convex economies. *Set-Valued Analysis*, 3:307–337, 1995.

[99] J. Greenberg. Quasi-equilibrium in abstract economies without ordered preferences. *J. Math. Econom.*, 4:163–165, 1977.

[100] B. Halpern. Fixed point theorems for set-valued maps in infinite dimensional spaces. *Math. Ann.*, 189:87–98, 1970.

[101] P. J. Hammond. Irreducibility, resource relatedness and survival in equilibrium with individual nonconvexities. In R. Becker, M. Boldrin, and W. Thomson, editors, *General Equilibrium, Growth and Trade*, pages 73–115. Academic Press, San Diego, 1993.

[102] O. D. Hart and H. W. Kuhn. A proof of the existence of equilibrium without the free disposal assumption. *J. Math. Econom.*, 2:335–343, 1975.

[103] W. Hildenbrand. Existence of equilibria for economies with production and a measure space of consumers. *Econometrica*, 38:608–623, 1970.

[104] W. Hildenbrand. *Core and equilibria in a large economy.* Princeton University Press, 1974.

[105] W. Hildenbrand and A. P. Kirman. *Introduction to Equilibrium Analysis.* North–Holland, Amsterdam and Oxford, 1976.

[106] C. J. Himmelberg. Fixed points of compact mutifunctions. *J. Math. Anal. Appl.*, 38:205–207, 1972.

[107] T. Ichiishi. On the Knaster–Kuratowski–Mazurkiewicz–Shapley theorem. *J. Math. Anal. Appl.*, 81:297–299, 1981.

[108] T. Ichiishi. *Game Theory for Economic Analysis.* Academic Press, New York and London, 1983.

[109] L. Jones. A competitive model of commodity differentiation. *Econometrica*, 52:507–530, 1984.

[110] L. Jones. Special problems arising in the study of economies with infinitely many commodities. In H. F. Sonnenschein, editor, *Models of Economic, Dynamics*, pages 184–205. Springer–Verlag, Berlin New York, 1986.

[111] L. Jones. Existence of equilibria with infinitely many commodities–Banach lattices reconsidered. *J. Math. Econom.*, 16:89–104, 1987.

[112] L. E. Jones. Existence of equilibria with infinitely many consumers and infinitely many commodities: a theorem based on models of commodity differentiation. *J. Math. Econom.*, 12:119–138, 1983.

[113] S. Kakutani. A generalization of Brouwer's fixed point theorem. *Duke Math. J.*, 8:457–459, 1941.

[114] J. L. Kelley. *General Topology*. D. Van Nostrand Company, New York, 1955.

[115] M. A. Khan. A remark on the existence of equilibria in markets without ordered preferences and with a Riesz space of commodities. *J. Math. Econom.*, 13:165–169, 1984.

[116] M. A. Khan and N. T. Peck. On the interiors of production sets in infinite dimensional spaces. *J. Math. Econom.*, 18:29–40, 1989.

[117] M. A. Khan and R. Vohra. On approximate decentralization of Pareto optimal allocations in locally convex spaces. *J. Approx. Theory*, 52:149–161, 1988.

[118] B. Knaster, K. Kuratowski, and S. Mazurkiewicz. Ein beweis des fixpunktsatzes für n-dimensionale simplexe. *Fund.Math.*, 14:132–137, 1929.

[119] H. W. Kuhn. A note on the law of supply and demand. *Math. Scand.*, 4:143–146, 1956.

[120] H. W. Kuhn. On a theorem of Wald. In H. W. Kuhn and A. W. Tucker, editors, *Linear Inequalities and Related Systems*, Ann. of Math. Stud. 38, pages 265–273. Princeton University Press, 1956.

[121] K. J. Lancaster. A new approach to consumer theory. *J. Political Economy*, LXXIV:132–157, 1966.

[122] K. J. Lancaster. Socially optimal product differentiation. *American Economic Review*, 65:567–585, 1975.

[123] J. M. Lasry and R. Robert. Degré et théorèmes de point fixe - applications. Technical report, Université de Paris-Dauphine, Paris, 1974.

[124] M. Magill. An equilibrium existence theorem. *J. Math. Anal. Appl.*, 84:162–169, 1981.

[125] V. M. Marakulin. Production equilibria in vector lattices: An approach using finite-dimensional approximations. Technical Report 9821, CEPREMAP, Paris, 1998.

[126] A. Mas-Colell. An equilibrium existence theorem without complete or transitive preferences. *J. Math. Econom.*, 1:237–246, 1974.

[127] A. Mas-Colell. A note on a theorem of F. Browder. *Math. Program.*, 6:229–233, 1974.

[128] A. Mas-Colell. A model of equilibrium with differentiated commodities. *J. Math. Econom.*, 2:263–296, 1975.

[129] A. Mas-Colell. Pareto optima and equilibria: The infinite dimensional case. In C. D. Aliprantis, O. Burkinshaw, and N. Rothman, editors, *Advances in Equilibrium Theory*, pages 25–42. Springer–Verlag, New York, 1985.

[130] A. Mas-Colell. The price equilibrium existence problem in topological vector lattices. *Econometrica*, 54:1039–1053, 1986.

[131] A. Mas-Colell. Valuation equilibrium and Pareto optimum revisited. In W. Hildenbrand and A. Mas-Colell, editors, *Contributions to Mathematical Economics*, pages 317–331. North–Holland, New York, 1986.

[132] A. Mas-Colell and S. Richard. A new approach to the existence of equilibria in vector lattices. *J. Econom. Theory*, 53:1–11, 1991.

[133] A. Mas-Colell and W. R. Zame. Equilibrium theory in infinite dimensional spaces. In W. Hildenbrand and H. Sonnenschein, editors, *Handbook of Mathematical Economics, Vol. 4*, pages 1836–1898. North–Holland, New York, 1991.

[134] L. McKenzie. On equilibrium in Graham's model of world trade and other competitive systems. *Econometrica*, 22:147–161, 1954.

[135] L. McKenzie. Competitive equilibrium with dependent consumer preferences. In H. A. Antosiewicz, editor, *Proceedings of the Second Symposium in Linear Programming*, pages 277–294. National Bureau of Standards, Washington DC, 1955.

[136] L. McKenzie. On the existence of general equilibrium for a competitive market. *Econometrica*, 27:54–71, 1959.

[137] L. McKenzie. On the existence of general equilibrium: some corrections. *Econometrica*, 29:247–248, 1961.

[138] L. McKenzie. The classical theorem on existence of competitive equilibrium. *Econometrica*, 49:819–841, 1981.

[139] E. Michael. Continuous selections I. *Annals of Mathematics*, 63:361–382, 1956.

[140] E. Michael. Continuous selections III. *Ann. of Math.*, 65:375–390, 1957.

[141] E. Michael. A survey of continuous selections. In A. Dold and B. Eckman, editors, *Set-valued mappings, selections and topological properties of* 2^X, Lecture Notes in Math.171, pages 54–58. Springer–Verlag, Berlin Heidelberg New York, 1970.

[142] J. Moore. The existence of 'compensated equilibrium' and the structure of the Pareto efficiency frontier. *Internat. Econom. Rev.*, 16:267–300, 1975.

[143] J. Nash. Equilibrium points in n-person games. *Proc. Natl.Acad. Sci. USA*, 36:48–49, 1950.

[144] T. Negishi. Welfare economics and existence of an equilibrium for a competitive economy. *Metroeconomica*, 12:92–97, 1960.

[145] H. Nikaido. On the classical multilateral exchange problem. *Metroeconomica*, 8:135–145, 1956.

[146] H. Nikaido. A supplementary note to Nikaido (1956). *Metroeconomica*, 9, 1957.

[147] H. Nikaido. *Convex Structures and Econom. Theory.* Academic Press, New York and London, 1968.

[148] B. Peleg and M. E. Yaari. Markets with countably many commodities. *Internat. Econom. Rev.*, 11:369–377, 1970.

[149] K. Podczeck. Equilibria in vector lattices without ordered preferences or uniform properness. *J. Math. Econ.*, 25:465–485, 1996.

[150] T. Rader. The existence of a utility function to represent preferences. *Review of Economic Studies*, 30:229–232, 1963.

[151] T. Rader. Edgeworth exchange and general economic equilibrium. *Yale Economic Essays*, 4:133–180, 1974.

[152] L. K. Raut. Myopic topologies on general commodity spaces. *J. Econom. Theory*, 39:358–367, 1986.

[153] S. F. Richard. Competitive equilibria in Riesz spaces. Technical report, GSIA, Carnegie Mellon University, 1986.

[154] S. F. Richard. A new approach to production equilibria in vector lattices. *J. Math. Econ.*, 18:41–56, 1989.

[155] S. F. Richard and W. R. Zame. Proper preferences and quasi-concave utility functions. *J. Math. Econom.*, 15:231–247, 1986.

[156] V. F. Martins Da Rocha. Equilibre général avec une double infinité d'agents et de biens. Technical report, Université de Paris 1, Paris, 2002. Thèse de doctorat de l'Université.

[157] R. T. Rockafellar. *Convex Analysis.* Princeton University Press, 1970.

[158] M. Rogalski. Surjectivité d'applications multivoques dans les convexes compacts. *Bull. Sc. Math.*, Deuxième série-96:83–87, 1972.

[159] H. Scarf. The core of an n person game. *Econometrica*, 35:50–69, 1967.

[160] H. Scarf. Notes on the core of a productive economy. In W. Hildenbrand and A. Mas-Colell, editors, *Contributions to Mathematical Economics, in honor of Gérard Debreu*, pages 400–429. North-Holland, Amsterdam, 1986.

[161] H. Schaefer. *Topological vectors spaces.* Macmillan, New York, 1966.

[162] J. Schauder. Der fixpunksatz in funktionalräuman. *Studia Math.*, 2:171–180, 1930.

[163] D. Schmeidler. Competitive equilibria in markets with a continuum of traders and incomplete preferences. *Econometrica*, 37:578–585, 1969.

[164] W. Shafer. The non-transitive consumer. *Econometrica*, 42:913–919, 1974.

[165] W. Shafer. Equilibrium in economies without ordered preferences or free-disposal. *J. Math. Econom.*, 3:135–137, 1976.

[166] W. Shafer and H. Sonnenschein. Equilibrium in abstract economies without ordered preferences. *J. Math. Econom.*, 2:345–348, 1975.

[167] W. Shafer and H. Sonnenschein. The non-transitive consumer. *J. Econom. Theory*, 11:83–93, 1975.

[168] L. S. Shapley. On balanced games without side payments. In T. C. Hu and S. M. Robinson, editors, *Math. Program.*, pages 261–290. Academic Press, New York, 1973.

[169] H. Sonnenschein. Demand theory without transitive preferences with applications to the theory of competitive equilibrium. In J. Chipman, L. Hurwicz, M. Richter, and H. Sonnenschein, editors, *Preferences, Utility and Demand.* Harcourt–Brace–Jovanovich, New York, 1971.

[170] F. Terkelsen. A short proof of Fan's fixed point theorem. *Proc. Amer. Math. Soc.*, 42:643–644, 1974.

[171] R. Tourky. A new approach to the limit theorem on the core of an economy in vector lattices. *J. Econom. Theory*, 78:321–328, 1998.

[172] R. Tourky. The limit theorem on the core of a production economy in vector lattices with unordered preferences. *Econom. Theory*, 14:219–226, 1999.

[173] S. Toussaint. On the existence of equilibria in economies with infinitely many commodities and without ordered preferences. *J. Econom. Theory*, 33:98–115, 1984.

[174] A. Tychonoff. Ein fixpunktsatz. *Math. Ann.*, 111:767–776, 1935.

[175] H. Uzawa. Walras existence theorem and Brouwer's fixed point theorem. *Economic Studies Quarterly*, 13 no. 1, 1962.

[176] A. Wald. Über die eindeutige positive lösbarkeit der neuen produktiongleichungen. *Ergebnisse eines Mathematischen Kolloquiums*, 6:12–20, 1935.

[177] A. Wald. Über die produktiongleichungen der ökononomischen wertlehre. *Ergebnisse eines Mathematischen Kolloquiums*, 7:1–6, 1936.

[178] A. Wald. Über einige gleichungssysteme der mathematischen ökonomie. *Zeitschrift für Nationalökonomie*, 7:637–670, 1936. Translated as "On some systems of equations of Mathematical Economics", *Econometrica* 19: 368–403, 1951.

[179] C. Weddepohl. Equilibrium in a market with incomplete preferences where the number of consumers may be finite. In G. Schwodiauer, editor, *Equilibrium and Disequilibrium in Economic Theory.* D. Reidel Publishing Company, Dordrecht, 1977.

[180] N. C. Yannelis and N. D. Prabhakar. Existence of maximal elements and equilibria in linear topological spaces. *J. Math. Econom.*, 13:233–245, 1983.

[181] N. C. Yannelis and W. R. Zame. Equilibria in Banach lattices without ordered preferences. *J. Math. Econom.*, 15:85–110, 1986.

[182] W. R. Zame. Competitive equilibria in production economies with an infinite-dimensional commodity space. *Econometrica*, 55:1075–1108, 1987.

Index of symbols

Chapter 1

Let X and Y be two sets.

A correspondence φ: $X \to Y$ is a function $X \mapsto 2^Y$.

For a correspondence φ: $X \to Y$,

$\mathcal{G}_\varphi \subset X \times Y$ denotes the graph of φ;

For $A \subset X$, $\varphi(A) := \bigcup_{x \in A} \varphi(x)$ denotes the image of A;

For $y \in Y$, $\varphi^{-1}(y) := \{x \in X\colon\ y \in \varphi(x)\}$ denotes the fiber of y.

Given a positive integer ℓ, $\mathbb{R}^\ell$ denotes the ℓ-dimensional Euclidean space;

$\mathbb{R}^\ell_+ :=$ canonical positive cone;

$\mathbb{R} := \mathbb{R}^1$;

$\mathbb{R}_+ := \mathbb{R}^1_+$;

For a real sequence $(\lambda_\nu)_{\nu \in \mathbb{N}}$,

$\lambda_\nu \downarrow 0$ means that the sequence is decreasing with infimum 0.

For any x, $y \in \mathbb{R}^\ell$, and $r > 0$ in $\mathbb{R}$,

$x_i :=$ the ith coordinate of x, $i = 1, \ldots, \ell$;

$x \cdot y := \sum_{i=1}^{\ell} x_i y_i$ denotes the Euclidean inner product of x and y;

$\|x\| := \sqrt{x \cdot x}$ denotes the Euclidean norm of x;

$x \geq y$ means $x_i \geq y_i$ for every $i = 1, \ldots, \ell$;

$x > y$ means $x \geq y$ and $x \neq y$;

$x \gg y$ means $x_i > y_i$ for every $i = 1, \ldots, \ell$;

$x^+ := x \vee 0 = (\max\{x_i, 0\})_{i=1}^{\ell}$ denotes the positive part of x;

$x^- := (-x) \vee 0 = (\max\{-x_i, 0\})_{i=1}^{\ell}$ denotes the negative part of x;

$B(x, r) := \{z \in \mathbb{R}^\ell\colon\ \|z - x\| < r\}$ denotes the open ball with center x and radius r;

$\overline{B}(x,r) := \{z \in \mathbb{R}^\ell \mid \|z-x\| \leq r\}$ denotes the closed ball with center x and radius r;
$B := \{x \in \mathbb{R}^\ell\colon\ \|x\| < 1\}$ denotes the (open) unit-ball of $\mathbb{R}^\ell$;
$\overline{B} := \{x \in \mathbb{R}^\ell\colon\ \|x\| \leq 1\}$ denotes the (closed) unit-ball of $\mathbb{R}^\ell$;
$S := \{x \in \mathbb{R}^\ell\colon\ \|x\| = 1\}$ denotes the unit-sphere of $\mathbb{R}^\ell$.

If X is a topological space, x a point of X and A a subset of X,
$\mathcal{V}(x)$ is the family of all neighborhoods of x;
$\mathcal{B}(x)$ is a local base of closed convex and circled neighborhoods of x;
$\operatorname{cl} A$ denotes the closure of A;
$\operatorname{int} A$ denotes the interior of A;
∂A denotes the boundary of A;
For $f\colon X \to \mathbb{R}_+$, $\operatorname{supp} f$ denotes $\operatorname{cl}\{x \in X\colon\ f(x) \neq 0\}$.

If X is a vector space, for any subsets A, B of X, for any $\lambda \in \mathbb{R}$,
$A + B := \{z \in X\colon\ z = x + y,\ x \in A,\ y \in B\}$;
$A - B := \{z \in X\colon\ z = x - y,\ x \in A,\ y \in B\}$;
$\lambda A := \{z \in X\colon\ z = \lambda x,\ \lambda \in \mathbb{R},\ x \in A\}$;
$\operatorname{co} A$ denotes the convex hull of A.

If X is a topological vector space,
$\overline{\operatorname{co}}\, A$ denotes the closed convex hull of A.

If $N = \{1, \ldots, n\}$ is a finite set and if $\mathcal{N} := 2^N \setminus \emptyset$, for $S \in \mathcal{N}$,
χ_S denotes the characteristic function of S;
$|S|$ denotes the number of elements of S.

Chapter 2

$\langle \mathbb{R}^\ell, \mathbb{R}^\ell \rangle$ is the commodity-price duality.
$\mathcal{E}$ denotes an economy. I is the set of consumers, J is the set of producers.
X_i denotes the consumption set of consumer i;
P_i is a strict preference correspondence;
$\widehat{P}_i$ denotes the augmented preference correspondence associated to P_i;
$\succsim_i$ is a preorder of preferences on X_i;
ω_i denotes the initial endowment of i;
ω denotes the total endowment;
Y_j denotes the production set of j;
$\theta_{i,j}$ denotes the profit share of i in j;
Y denotes the total production set;
Z is a disposal cone;
$(x, y) \in \prod_{i \in I} X_i \times \prod_{j \in J} Y_j$ is an allocation;

$\mathcal{A}(\mathcal{E})$ is the set of all attainable allocations;
$\widehat{X}$ is the set of all attainable consumption allocations;
$\widehat{Y}$ is the attainable total production set;
$\widehat{X}_i$, $\widehat{Y}_j$ are the attainable sets of consumer i and producer j.

For a subset C of $\mathbb{R}^\ell$,
$\mathrm{A}(C)$ denotes the recession cone of C;
If $c \in \operatorname{cl} C$, $T_c(C)$ denotes the tangent cone to C at c.

Chapter 3

Γ is an abstract economy. N is a set of agents.
If Γ is an abstract economy,
X_i denotes the strategy set of agent i;
$\alpha_i \colon \prod_{k\in I} X_k \to X_i$ denotes the constraint correspondence of i;
$P_i \colon \prod_{k\in I} X_k \to X_i$ denotes the strict preference correspondence of i;
$\mathcal{A}(\Gamma)$ denotes the set of feasible strategy vectors.

Chapter 4

I is the set of consumers; $S \subset I$, $S \neq \emptyset$ denotes a coalition.
For a coalition S,
$\widehat{X}_S$ denotes the set of all attainable consumption assignments for S,
P_S denotes the preference correspondence of S;
$\mathcal{C}(\mathcal{E})$ denotes the core of $\mathcal{E}$;
If r is a positive integer,
$\mathcal{E}^r$ denotes the r-replica of $\mathcal{E}$;
$\mathcal{C}^r(\mathcal{E})$ denotes the set of all attainable allocations of $\mathcal{E}$ whose r-replica belongs to $\mathcal{C}(\mathcal{E}^r)$;
$\mathcal{C}^e(\mathcal{E}) = \bigcap_{r\geq 1} \mathcal{C}^r(\mathcal{E})$ denotes the set of Edgeworth equilibria;
$t = (t_i)_{i\in I} \in [0,1]^I \setminus \{0\}$ denotes a fuzzy coalition;
$\mathcal{T}$ denotes the set of all fuzzy coalitions;
Y_t denotes the production possibility set of the fuzzy coalition t;
$\widehat{X}_t$ denotes the set of all attainable consumption assignments for t;
$\mathcal{C}^f(\mathcal{E})$ denotes the fuzzy core of $\mathcal{E}$.

Chapter 5

If (Ω, Σ, μ) is a measure space,
$L_\infty(\Omega, \Sigma, \mu)$ denotes the space of (equivalence classes of) μ-measurable, μ-essentially bounded functions $(\Omega, \Sigma, \mu) \mapsto \mathbb{R}$;

$L^\ell_\infty(\Omega, \Sigma, \mu)$ denotes the space of (equivalence classes of) μ-measurable, μ-essentially bounded functions $(\Omega, \Sigma, \mu) \mapsto \mathbb{R}^\ell$.

For $1 \leq p < \infty$, $L_p(\Omega, \Sigma, \mu)$ denotes the space of (equivalence classes of) μ-measurable, p-integrable functions $(\Omega, \Sigma, \mu) \mapsto \mathbb{R}$;

$L^\ell_p(\Omega, \Sigma, \mu)$ denotes the space of (equivalence classes of) μ-measurable, p-integrable functions $(\Omega, \Sigma, \mu) \mapsto \mathbb{R}^\ell$.

If K is a compact metric space,

$M(K)$ denotes the space of all countably additive signed measures on K;

$C(K)$ is the set of continuous real functions defined on K.

If L is a vector space,

L^* denotes the algebraic dual of L.

Let L and L' be two vector spaces in separating duality.

$\langle L, L' \rangle$ denotes a commodity–price duality for the economy $\mathcal{E}$;

$(x, p) \mapsto p \cdot x = \langle x, p \rangle$ denotes the evaluation function;

If $A \subset L$, $p \cdot A := \{p \cdot x\colon\ x \in A\}$;

σ, τ denote diverse vector space topologies on L, compatible or not with the duality;

σ^I denotes the product topology on $(L, \sigma)^I$;

$\sigma(L, L')$ denotes the weak topology on L;

$\tau(L, L')$ denotes the Mackey topology on L relative to the duality;

$\sigma(L', L)$ denotes the weak-star topology on L'.

If Z is a cone of Z, Z^0 denotes the polar cone of Z in L';

If $A \subset L$, $A^0 = \{p \in L'\colon\ p \cdot x \leq 1 \text{ for each } x \in A\}$ denotes the polar set of A in L'.

If F is a finite dimensional vector subspace of L,

$\mathcal{E}_F$ denotes the economy $\mathcal{E}$ restricted to F;

If K is a vector subspace of L,

$\mathcal{E}_K$ denotes the economy $\mathcal{E}$ restricted to K.

For a topology τ on L,

$\operatorname{int} A$ denotes the τ-interior of the set $A \subset L$;

$\operatorname{cl} A$ denotes the τ-closure of A;

$(L, \tau)'$ denotes the topological dual of (L, τ).

If L is an ordered vector space,

L_+ denotes the positive cone.

If L is a vector lattice, for any x, $y \in L$,

$x \vee y$ denotes the supremum of x and y;

$x \wedge y$ denotes the infimum of x and y;
$x^+ := x \vee 0$ denotes the positive part of x;
$x^- := (-x) \vee 0$ denotes the negative part of x
For $a, b \in L$, $a \leq b$, $[a, b] := \{x \in L\colon\ a \leq x \leq b\}$.

$L^\sim$ denotes the order dual of the vector lattice L, and for $f, g \in L^\sim$,
$f \vee g$ denotes the supremum of f and g;
$f \wedge g$ denotes the infimum of f and g.
For a finite list $f = (f_1, \ldots f_n)$ of linear functionals in $L^\sim$,
$\bigvee_{i=1}^n f_i$ denotes the supremum of $f_1, \ldots, f_n$ and for $x \geq 0$,
$\mathcal{R}_f(x)$ denotes the value at x of this supremum.

Appendix

For a correspondence $\varphi\colon X \to Y$, if $A \subset Y$,
$\varphi^+(A) = \{x \in X\ :\ \varphi(x) \subset A\}$, (superior inverse);
$\varphi^-(A) = \{x \in X\ :\ \varphi(x) \cap A \neq \emptyset\}$, (inferior inverse);
$\mathcal{G}_\varphi \subset X \times Y$ denotes the graph of φ.

If V is a subset of X, V^c denotes the complementary of V in X.

Index